# Microwave Technology

***The Artech House Microwave Library***

*Introduction to Microwaves* by Fred E. Gardiol
*Receiving Systems Design* by Stephen J. Erst
*Applications of GaAs MESFETs* by R.A. Soares, J. Graffeuil, and J. Obregon
*GaAs Processing Techniques* by R.E. Williams
*GaAs FET Principles and Technology,* J.V. DiLorenzo and D. Khandelwal, eds.
*Modern Spectrum Analyzer Theory and Applications* by Morris Engelson
*Microwave Materials and Fabrication Techniques* by Thomas S. Laverghetta
*Handbook of Microwave Testing* by Thomas S. Laverghetta
*Microwave Measurements and Techniques* by Thomas S. Laverghetta
*Principles of Electromagnetic Compatibility* by Bernhard E. Keiser
*Microwave Filters, Impedance Matching Networks, and Coupling Structures* by G.L. Matthaei, Leo Young, and E.M.T. Jones
*Microwave Engineer's Handbook*, 2 vol., Theodore Saad, ed.
*Computer-Aided Design of Microwave Circuits* by K.C. Gupta, R. Garg, and R. Chadha
*Microstrip Lines and Slotlines* by K.C. Gupta, R. Garg, and I.J. Bahl
*Microstrip Antennas* by I.J. Bahl and P. Bhartia
*Microwave Circuit Design Using Programmable Calculators* by J. Lamar Allen and Max Medley, Jr.
*Stripline Circuit Design* by Harlan Howe, Jr.
*Microwave Transmission Line Filters* by J.A.G. Malherbe
*Electrical Characteristics of Transmission Lines* by W. Hilberg
*Multiconductor Transmission Line Analysis* by Sidney Frankel
*Microwave Diode Control Devices* by Robert V. Garver
*A Practical Introduction to Impedance Matching* by Robert Thomas
*Active Filter Design* by A.B. Williams
*Adaptive Electronics* by Wolfgang Gaertner
*Laser Applications* by W.V. Smith
*Electronic Information Processing* by W.V. Smith
*Logarithmic Video Amplifiers* by Richard S. Hughes
*Avalanche Transit-Time Devices*, George Haddad, ed.
*Gallium Arsenide Bulk and Transit-Time Devices*, Lester Eastman, ed.
*Ferrite Control Components*, 2 vol., Lawrence Whicker, ed.

# Microwave Technology

Erich
Pehl

**International Standard Book Number: 0-89006-164-5**
**Library of Congress Catalog Card Number: 85-70815**

A translation from the German of *Mikrowellentechnik: Wellenleitungen und Leitungsbausteine,* © 1983 Dr. Alfred Huthig Verlag, Heidelberg, Federal Republic of Germany.

## ■ CONTENTS

# FOREWORD

In this book, *Microwave Technology,* we will be dealing with electromagnetic waves, waveguides which direct these waves (TEM-circuits, hollow waveguides, microstrip circuits, dielectric circuits), and circuit components.

An additional volume published in German deals with microwave antennas, microwave tubes, microwave semiconductors (diodes, bipolar and field-effect transistors), and semiconductor circuits such as mixer tubes, oscillators, and amplifiers.

This book addresses itself to students of electrical engineering and physics, and to the reader who is professionally or personally interested in microwave technology as a means of familiarization or as a reference work.

I wish to thank warmly the publishers, Dr. Alfred Hüthig Verlag, for their cooperation, their obliging compliance with my wishes, and the fine layout of the book. My special thanks go to the chief editor, Mr. Curt Rint.

*Erich Pehl*
*Osnabruck, Autumn 1983*

# INTRODUCTION

We usually designate microwaves as those electromagnetic waves within the frequency range of 300 MHz to 300 GHz, with respective wavelengths $\lambda = c/f$ from one meter to one millimeter.

The wavelengths are therefore, approximately on the same order of magnitude as the size of the circuit components which are used. When the lengths of the circuit components at high frequencies become comparable to the wavelengths used, the electrical tension and current, or rather the electrical and magnetic field intensities, yield various values at different positions of the component (for example, a circuit) at a fixed point in time and there is a phase shift of these lengths between neighboring points. In addition to time, the signals also depend upon position; i.e., they possess a wave character which cannot be neglected, as was the case in the low frequency range. On the other hand, the wavelengths are still not small enough to use only the methods of geometric optics.

The wave character of signals whose wavelengths are comparable to the magnitude of the circuit components leads to unique principles, in contrast to direct-current or alternating-current engineering, for the realization of structural components. Therefore, for example, sections of the circuit can be used in the construction of capacitor, inductors, and oscillator circuits; or frequency modulation effects in electron tubes or semiconductor elements can be used in the generation or amplification of oscillations. From the perspective of low-frequency engineering, another approach is necessary, one that makes use of the wave concept and which must frequently consider the high-frequency electrical and magnetic fields instead of the integral magnitudes of the tension and current.

Microwave technology has a wide range of applications and the development of microwave semiconductor components has created new areas of application. The use of higher and higher frequencies is contingent on the heavy utilization of the lower frequency bands and the increasing need for new frequency channels. This consideration aside, the utilization of higher frequencies has some advantages: in the case of a communications transmission with higher carrier frequencies, relatively smaller frequency ranges become available. At high frequencies, directional antennas with greater gain and a strong beam effect of relatively small size can be constructed. At higher frequencies, one obtains better resolution for radar instruments and, with circuit construction which uses the Doppler principle, a greater Doppler frequency distortion.

The fields of microwave application include: radio communication technology, remote sensing and satellite communication engineering; radar technology and direction finding for navigation and telemetry; the possibility of high-speed data transmission and processing; radio astronomy, field protective devices, and safety radars for traffic, telemetry, and remote control (for guidance vehicles); dielectrical heating for warming meals, and drying paper, wood, and textiles; welding synthetics or diathermy instruments in medicine; investigation of material characteristics; and material testing, gas spectrometry, and production monitoring.

In the first chapter of this book the propagation of electromagnetic energy in free space in the form of surface waves is explained first. Next, we will demonstrate which wave forms on conductor surfaces (which serve the direction of waves) are possible, subject to the limit conditions which must fill the electrical and magnetic fields. In connection with this, we will observe circuits that direct waves, and on which the electrical field and the magnetic field (in an ideal conductor) run absolutely transverse to the direction of propagation (L-waves on two-wire layouts, as, for example, the parallel-wire line and the coaxial transmission line), and circuits on which either the magnetic or the electrical field also possess longitudinal components (H-waves or E-waves on hollow waveguides). We will then deal with various types of microstrip circuits and, finally, with dielectric circuits upon which the energy propagates in the form of surface waves.

After the study of straggling parameters (S-parameters), which serve to illustrate microwave circuit arrangements with the help of power waves (Chapter 2), Chapters 3 and 4 will introduce microwave components in waveguide engineering and strip-line technology, such as resonators, matching components, electric-wave filters, attenuator pads, transfer circuits, and bypasses, and directional couplers. Microwave circuit arrangements can be realized in a small amount of space with the help of strip-line technology and in connection with semiconductor components. In this connection, various circuit types are also specified which are suitable for the construction of integrated circuit arrangements in the millimeter-wave range.

Components that work with microwave ferrites and which primarily serve the realization of non-reciprocal transmission behavior (one-way circuits, circulators) will be described in Chapter 5.

Throughout this book, such physical fundamentals as may from time to time be necessary will be provided.

## CHAPTER 1

# ELECTROMAGNETIC WAVES

## 1.1 THE CONCEPT OF WAVES IN MECHANICS

We will first examine the concept of a wave throgh an illustrative example free from mechanics: a wave on a straight line or waves in the water.

If one throws a stone into still water, the water will be moved downward, at the striking point, which, because of the repelling forces, causes approximately vertical harmonic oscillations. As a result of the elastic coupling between neighboring water particles, these too will be generated as dephased vertical oscillations. The force at the striking point will be pushed outward. Figure 1.1 represents two consecutive moments of a cross section of a disturbed water surface or an oscillating line (without attenuation and reflection).

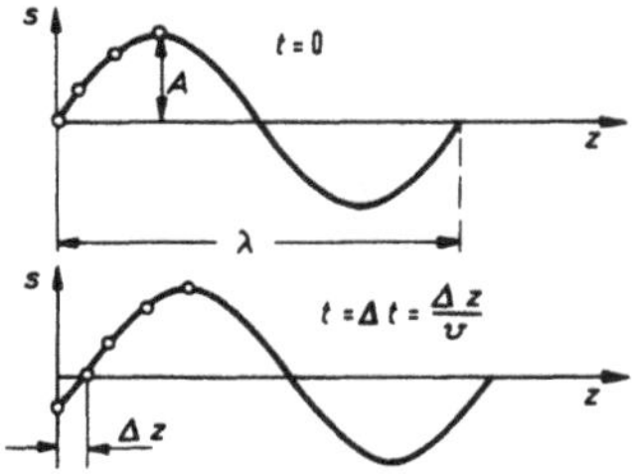


**Fig. 1.1** Mechanical transverse wave

We can see that a very convenient neighboring particle lags in phase, and that it does so proportionally to the distance to the reference particle. Therefore, the oscillation phase shifts to the right, and a wave propagates to the right in the $z$-direction. After one full period of oscillation $t = T$ of a particle, we obtain again the same picture as for $t = 0$. We designate the shortest distance between two particles of the same oscillation phase as wavelength $\lambda$. Therefore, the propagation speed of an oscillation phase is

$$v = \frac{\lambda}{T}\,. \tag{1.1}$$

The phase lag $\delta\varphi$ of a random particle at point $z$ results from the proportion $\Delta\varphi = 2\pi$ for $z = \lambda$ to $\Delta\varphi = 2\pi/\lambda \cdot z$. Therefore, the deflection $s$ of a particle at point $z$ and at time $t$ on a wave propagating in the positive $z$-direction can be stated as

$$S = A \cdot \sin(\omega t - \Delta\varphi) = A \cdot \sin(\omega t - 2\pi/\lambda \cdot z) \,. \tag{1.2}$$

Here $A$ is the amplitude of the wave and $\omega = 2\pi f = 2\pi/T$ its cyclic frequency. As an initial condition, we assumed $s = 0$ for $t = 0$. Since in the case just observed the vibration direction of the particles follows perpendicularly to the propagation direction of the wave, we may speak here of a cross wave or transverse wave.

## 1.2 FIELD EQUATIONS

Electromagnetic waves can be understood as a coupling of electrical and magnetic alternating fields.

Moving charge carriers in a conductor (as, for example, an ionic current) create a magnetic field which circularly encloses the current path. If the path is interrupted at a point (by a condenser, for example [Fig. 1.2]), then the alternating electrical field present at the point of interruption also surrounds itself with a magnetic field. We can conceive that the conductor current continues as a shift current in the dielectric (introduced by J. Maxwell).

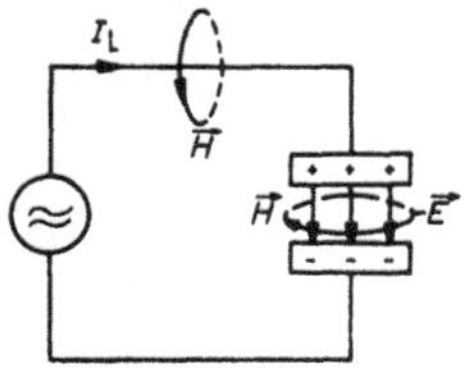


**Fig. 1.2** Explanation of the circulation theorem

The mathematical formulation of this relationship is the *circulation theorem* (A.M. Ampere, 1826)

$$\oint \vec{H} \cdot d\vec{s} = i_L + i_V = \int \vec{S}_L \cdot d\vec{A} + \int \frac{d\vec{D}}{dt}\, d\vec{A} = \int \varkappa \vec{E} \cdot d\vec{A} + \int \varepsilon \frac{d\vec{E}}{dt} \cdot d\vec{A} \,. \tag{1.3}$$

The integral of the magnetic field intensity $\vec{H}$ over a closed path $s$ is equal to the conductor current $i_L$ and the shift current $i_V$, which fill the surface created along this path. The conductor current and the shift current can be expressed by the integration area of the conductor current density $S_L$ and the shift current density $dD/dt$.

According to the law of induction (Faraday's law), a magnetic field fluctuating in time surrounds itself with a circular electrical field. This circular electrical field causes an induced electrical voltage in a conductor whose components lie in the direction of conductance. If the conductor creates a closed short-circuit which is permeated by a varying magnetic field, then the electromotive force that is actively induced by this short-circuit creates a secondary magnetic field counteracting its original cause (Lenz's rule).

The law of induction (M. Faraday, 1831) can be mathematically formulated as

$$\int \vec{E} \cdot d\vec{s} = -\frac{d\Phi}{dt} \tag{1.4}$$

with the magnetic flux

$$\Phi = \int \vec{B} \cdot d\vec{A} = \int \mu \vec{H} \cdot d\vec{A} \, . \tag{1.5}$$

The contour integral of the electrical field intensity $\vec{E}$ or the induced electromotive force is proportional to the time change of the magnetic flux which flows through this circular path.

Thus, with fields fluctuating in time, the electrical field intensity creates a magnetic field over the conductor current and the shift current, creating once again an electrical field through the induction effect. This consideration led Maxwell (1864) to assume the existence of electromagnetic waves, for which the experimental proof was given twenty-three years later by Heinrich Hertz.

The relationships of Eqs. (1.3) and (1.4) with Eq. (1.5), which are represented in integral form, can in many cases be much more easily evaluated by the differential display chosen by Maxwell. Thus, according to Eq. (1.3), we obtain the current density $S$ at a given point by solving the closed line integral for the limit of a surface element at a position perpendicular to the current density vector, whereby the magnitude of the surface element is allowed to shrink to zero:

$$S = \lim_{\Delta A \to 0} \frac{1}{\Delta A} \oint_{(C)} \vec{H} \cdot d\vec{s} = |\text{curl} \quad \vec{H}| \, . \tag{1.6}$$

The operation executed with vector $\vec{H}$, according to Eq. (1.6), yields the magnitude of a vector designated as an $\vec{H}$ rotation (turbulence). The direction of the vector curl $\vec{H}$ reveals itself as the direction perpendicular to the surface element $dA$, which, in the right-handed sense, is associated with the rotation direction of the line integral (compare with Fig. 1.3a).

Therefore, Eqs. (1.3) and (1.4) read with Eq. (1.5) in differential notation as

$$\text{curl}\ \vec{H} = \vec{S}_L + \frac{d\vec{D}}{dt} = \varkappa \cdot \vec{E} + \varepsilon \cdot \frac{d\vec{E}}{dt} \tag{1.7}$$

$$\text{curl}\ \vec{E} = -\frac{d\vec{B}}{dt} = -\mu \frac{d\vec{H}}{dt} \, . \tag{1.8}$$

Equations (1.7) and (1.8) are also designated as Maxwell's first and second equations.

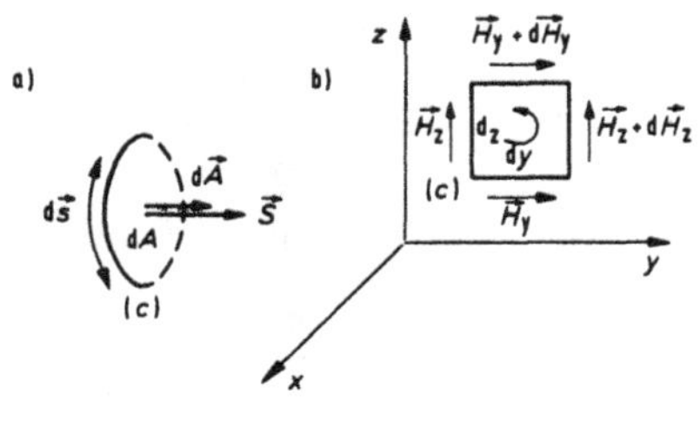


**Fig. 1.3** Formation of rotation

In this connection, the so-called constitutive equations are valid:

$$\vec{D} = \varepsilon_r \varepsilon_0 \vec{E}\ , \quad \vec{B} = \mu_r \mu_0 \vec{H}\ , \quad \vec{S}_L = \varkappa \cdot \vec{E}\ . \tag{1.9}$$

The formation of the rotational components of a vector in Cartesian coordinates is derived for the $x$-components.

One must choose a surface element $dA = dy \cdot dz$ perpendicular to the $x$-direction. Figure 1.3b shows the corresponding surface of a differential element of volume in the area under consideration; the rotation direction, which must be chosen by solving the closed line integral $\oint \vec{H} \cdot d\vec{s}$, and the field components taken into consideration for the general case of an inhomogeneous field are entered there. If we express the differential field parts $d\vec{H}_y$ and $d\vec{H}_z$ through the complete differentials, we obtain

$$\oint_{(C)} \vec{H} \cdot d\vec{s} = H_y \cdot dy + \left(H_z + \frac{\partial H_z}{\partial y}\, dy\right) dz - \left(H_y + \frac{\partial H_y}{\partial z}\, dz\right) dy - H_z \cdot dz =$$

$$= dy\, dz \left(\frac{\partial H_z}{\partial y} - \frac{\partial H_y}{\partial z}\right) . \tag{1.10}$$

Therefore, we obtain for the $x$-components of the vector curl $\vec{H}$

$$\vec{e}_x \cdot \mathrm{rot}\, \vec{H} = \frac{\oint_{(C)} \vec{H} \cdot d\vec{s}}{dy \cdot dz} = \frac{\partial H_z}{\partial y} - \frac{\partial H_y}{\partial z}\ . \tag{1.11}$$

A consideration made on the basis of differential surface elements in both other basic planes yields

$$\vec{e}_y \cdot \text{curl } \vec{H} = \frac{\partial H_x}{\partial z} - \frac{\partial H_z}{\partial x} \tag{1.12}$$

$$\vec{e}_z \cdot \text{curl } \vec{H} = \frac{\partial H_y}{\partial x} - \frac{\partial H_x}{\partial y} . \tag{1.13}$$

On the whole, the vector curl $\vec{H}$ can be represented by the determinants

$$\text{curl } \vec{H} = \begin{vmatrix} \vec{e}_x & \vec{e}_y & \vec{e}_z \\ \frac{\partial}{\partial x} & \frac{\partial}{\partial y} & \frac{\partial}{\partial z} \\ H_x & H_y & H_z \end{vmatrix} . \tag{1.14}$$

The two Maxwellian field equations (1.7) and (1.8) are supplemented by additional conditions on the sources of the fields.

The sources of the electrical fields are the electrical charges. The charge quantity $\Delta Q$ present in an element of volume $\Delta V$ can be determined by the integration of the electrical shift density $\vec{D}$ over the surface $\Delta O$ of this element of volume:

$$\Delta Q = \oint_{(\Delta O)} \vec{D} \cdot d\vec{A} . \tag{1.15}$$

Through the formation of a high-pass filter $\Delta V \to 0$ we obtain the so-called space charge density for the position in question:

$$q_v = \lim_{\Delta V \to 0} \frac{\Delta Q}{\Delta V} . \tag{1.16}$$

We designate the solution of this high-pass filter in Eq. (1.15) as the divergence of vector $\vec{D}$. It is valid for

$$\lim_{\Delta V \to 0} \frac{1}{\Delta V} \int_{(\Delta O)} \vec{D} \cdot d\vec{A} \equiv \text{div } \vec{D} = q_v . \tag{1.17}$$

Divergence is a criterion for the flux current drawn out from an element of volume (source) and for the current flowing in (negative source). Since the magnetic field is solenoidal, an additional condition is valid:

$$\operatorname{div} \vec{B} = 0 . \tag{1.18}$$

The continuity of the conductor current and the shift current is expressed in the continuity equation of the electrical current as

$$\operatorname{div} \left( \vec{S}_L + \frac{\partial \vec{D}}{\partial t} \right) = 0 . \tag{1.19}$$

Equations (1.7), (1.8), (1.9) and (1.17), (1.18), (1.19) form the complete system of field equations necessary for solving field problems.

The divergence of a vector can be derived from an element of volume $dV = dx \cdot dy \cdot dz$ in Cartesian coordinates:

$$\operatorname{div} \vec{F} = \frac{\partial F_x}{\partial x} + \frac{\partial F_y}{\partial y} + \frac{\partial F_z}{\partial z} . \tag{1.20}$$

## 1.3 PLANE WAVES

We will investigate the theoretical case of the propagation of a wave in an unlimited and solenoidal medium, whose material characteristics are independent of location and direction (homogeneous and isotropic). The field magnitudes change in only one direction; unless this is the $z$-direction, in which case the following holds true:

$$\frac{\partial}{\partial x} = \frac{\partial}{\partial y} = 0 . \tag{1.21}$$

The time dependence of the field magnitudes is sinusoidal (steady state). Through a complex notation with the time factor $e^{j\omega t}$ we obtain the time derivation of field strength with the multiplication of $j\omega$:

$$\underline{E} = \hat{E} \cdot e^{(j\omega t + \varphi)} \left( \frac{d\underline{E}}{dt} = j\omega \cdot \hat{E} \cdot e^{j(\omega t + \varphi)} = j\omega \cdot \underline{E} \right) .$$

Maxwell's first equation (1.7) yields the following correlations:

$$\frac{\partial \underline{H}_y}{\partial z} = -(\varkappa + j\omega\varepsilon) \cdot \underline{E}_x , \tag{1.22}$$

$$\frac{\partial \underline{H}_x}{\partial z} = (\varkappa + j\omega\varepsilon) \cdot \underline{E}_y , \tag{1.23}$$

$$\underline{E}_z = 0 \tag{1.24}$$

and Maxwell's second equation (1.8)

$$\frac{\partial \underline{E}_x}{\partial z} = -\,j\omega\mu \underline{H}_y\,, \tag{1.25}$$

$$\frac{\partial \underline{E}_y}{\partial z} = j\omega\mu \underline{H}_x\,, \tag{1.26}$$

$$\underline{H}_z = 0\,. \tag{1.27}$$

We obtain two independent sets of equations: (1.22), (1.25) for the field intensity combinations $\underline{E}_x$, $\underline{H}_y$, and Eqs. (1.23), (1.26) for the pair $\underline{E}_y$, $\underline{H}_x$.

We will first examine the case in which the field would display the components $\underline{E}_x$ and $\underline{H}_y$.

After differentiating Eq. (1.23) and substituting in Eq. (1.22), we obtain the differential equations

$$\frac{\partial^2 \underline{E}_x}{\partial z^2} + (\omega^2\mu\varepsilon - j\omega\mu\varkappa)\,\underline{E}_x = 0 \tag{1.28}$$

respectively,

$$\frac{\partial^2 \underline{E}_x}{\partial z^2} - \gamma^2 \underline{E}_x = 0 \tag{1.29}$$

with

$$\gamma^2 = j\omega\mu\varkappa - \omega^2\mu\varepsilon\,. \tag{1.30}$$

The solution of the differential Eq. (1.29) is

$$\underline{E}_x(z) = \underline{E}_{x,h}\cdot e^{-\gamma z} + \underline{E}_{x,r}\cdot e^{\gamma z}\,. \tag{1.31}$$

$$\underline{H}_y(z) = -\frac{j\gamma}{\omega\mu}\left(\underline{E}_{x,h}\cdot e^{-\gamma z} - \underline{E}_{x,r}\cdot e^{\gamma z}\right) = \underline{H}_{y,h}\cdot e^{-\gamma z} + \underline{H}_{y,r}\cdot e^{\gamma z}\,. \tag{1.32}$$

Equations (1.31), (1.32) describe the overlapping of an electromagnetic wave propagating in the $z$-direction (index h for a forward moving wave) with a wave propagating in the opposite $z$-direction (index r for a receding or reflected wave).

The value $\gamma$ according to Eq. (1.30) is

$$\gamma = \alpha + j\beta = \sqrt{j\omega\mu\varkappa - \omega^2\mu\varepsilon} = \sqrt{j\omega\mu\,(\varkappa + j\omega\varepsilon)} \tag{1.33}$$

which is designed as the *propagation constant* or the *coefficient of transfer constant.* The *attenuation constant* $\alpha$ specifies the reduction of amplitude according to length in the direction of propagation, and the *phase constant* $\beta$ specifies the phase change in the direction of propagation of a wave.

The time dependence of field strength is not explicitly expressed in Eqs. (1.31) and (1.32). It can be represented through multiplication with the complex time factor $e^{j\omega t}$.

In real notation, the electrical field intensity of the forward moving wave, for example, is represented according to the structure of its real part ($\underline{E}_{x,h}$ being as

$$E_x(z, t) = \mathrm{Re}\,\{\underline{E}_{x,h} \cdot e^{-\gamma z} \cdot e^{j\omega t}\} = \hat{E}_{x,h} \cdot \cos(\omega t - \beta z) \,. \qquad (1.34)$$

We obtain a relationship corresponding to that of Eq. (1.2), a mechanical transverse wave propagating in the $z$-direction. Such an electromagnetic wave propagating in a given direction, whose electrical and magnetic field intensity vectors are perpendicular to each other and perpendicular to the direction of propagation, and which do not change in planes perpendicular to the direction of propagation, is called a *plane wave.* Surfaces with constant oscillation phase of their field intensity vectors, designated as wave fronts or phase fronts, create planes perpendicular to the direction of propagation (normal wave direction).

The plane wave can be used as a model for many advancing wave fields. The field of a random beam source can be regarded partly as a plane wave at a correspondingly great distance from the source in an undisturbed area.

The vectorial field intensities of the wave oscillate in a fixed direction in space; in this case, we speak of a *linear polarization* of the wave. We specify the direction of the electrical field intensity vector as the direction of polarization (therefore, in the above example, the $x$-direction).

An electromagnetic wave whose electrical and magnetic fields are transverse (directed perpendicular to the direction of propagation), is also called a *transverse electromagnetic* (TEM) *wave.*

The distance between two neighboring planes of equal oscillation phase of a wave propagating in one direction is designated as wavelength $\lambda$. With the condition $\beta \cdot \lambda = 2\pi$ [compare, for example, with Eq. (1.34)], the following results:

$$\lambda = \frac{2\pi}{\beta} \qquad (1.35)$$

($2\pi/\lambda$ is also designated as propagation factor $k$).

From Eq. (1.32) we obtain the relationship, which is independent of location, of the complex amplitudes of the electrical and magnetic field intensities at a given point of the wave field for the advancing or receding wave

$$\underline{Z}_F = \frac{\underline{E}_{x,h}}{\underline{H}_{y,h}} = -\frac{\underline{E}_{x,r}}{\underline{H}_{y,r}} = \frac{j\omega\mu}{\gamma} = \sqrt{\frac{j\omega\mu}{\varkappa + j\omega\varepsilon}} \,. \qquad (1.36)$$

In the case of waveguides with an unobstructed wave path, the relationship of the complex amplitudes of the electrical and magnetic cross field intensities at the same point in the field is defined as the *field impedance.* If we are dealing with the field amplitudes of a propagating wave – as in Eq. (1.36) – then the field impedance is designated as the *characteristic wave impedance* $\underline{Z}_F$.

If the propagation medium is non-dissipative ($\kappa = 0$, $\epsilon$ and $\mu$ real), then the propagation constant is a purely imaginary $\gamma = j\beta$

$$\alpha = 0, \quad \beta = \frac{2\pi}{\lambda} = \omega\sqrt{\mu\varepsilon}\,. \tag{1.37}$$

The wave is not attenuated and the phase constant undergoes a linear increase with frequency.

The characteristic wave impedance becomes real

$$Z_F = \sqrt{\frac{\mu}{\varepsilon}}\,. \tag{1.38}$$

This means that the electrical and magnetic field intensities are oscillating in phase.

The characteristic wave impedance in a vacuum yields

$$Z_F = \sqrt{\frac{\mu_0}{\varepsilon_0}} = 120\pi \cdot \Omega \approx 377\ \Omega\,. \tag{1.39}$$

Fig. 1.4 illustrates the path of the field intensity components dependent on $z$ of a plane wave propagating in the $z$-direction through a non-dissipative medium, and the lines of energy at a fixed point in time.

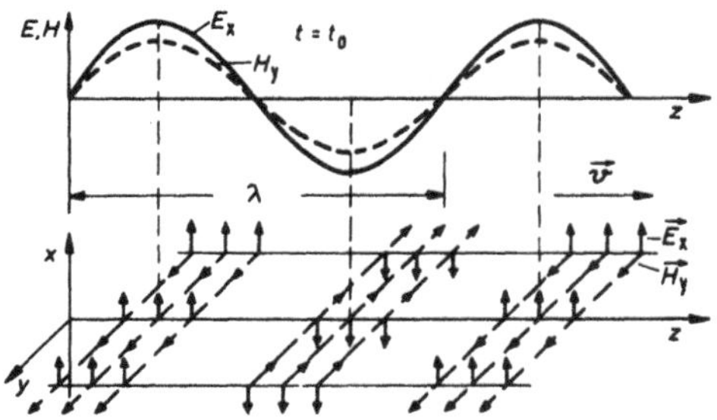


**Fig. 1.4** Field distribution of a plane wave in a non-dissipating medium

The graph of the field distribution of a wave propagating in one direction shifts in this direction at a speed designated as the *phase velocity* $v_p$. For this velocity, at which a certain phase condition propagates (the phase remains constant for an observer who is also moving), $\varphi = \omega t - \beta z = \text{const}$ [compare with Eq. (1.34)] is yielded upon differentiation of $\omega \cdot dt - \beta \cdot dz = 0$

$$v_p = \frac{dz}{dt} = \frac{\omega}{\beta}\,. \tag{1.40}$$

For non-dissipative wave propagation, we obtain by Eq. (1.37)

$$v_p = \frac{1}{\sqrt{\mu_0\mu_r \cdot \varepsilon_0\varepsilon_r}} = \frac{c}{\sqrt{\mu_r \cdot \varepsilon_r}} \tag{1.41}$$

and in free space with $\mu_r = \epsilon_r = 1$, the phase velocity is equal to the speed of light $c$.

For the group velocity $v_g$, at which the signal of a modulated wave from a small frequency group* or the energy of an advancing wave propagates, the following is valid:

$$v_g = \frac{d\omega}{d\beta} \,. \tag{1.42}$$

For the wavelengths of a non-dissipating advancing wave, we obtain from Eqs. (1.35), (1.37), (1.41)

$$\lambda = \frac{2\pi}{\beta} = \frac{2\pi}{\omega\sqrt{\mu\varepsilon}} = \frac{1}{f \cdot \sqrt{\mu\varepsilon}} = \frac{v_p}{f} = \frac{\lambda_0}{\sqrt{\mu_r \cdot \varepsilon_r}} \tag{1.43}$$

with $\lambda_0 = c/f$ = the wavelength in free space.

By solving Maxwell's equations in the case of plane waves, we obtain yet a second possible set of equations [Eqs. (1.23), (1.26)] for the field intensity combination $\underline{E}_y, \underline{H}_x$. According to a similar rule of calculation, we now obtain for the field components

$$\underline{E}_y(z) = \underline{E}_{y,h} \cdot e^{-\gamma z} + \underline{E}_{y,r} \cdot e^{\gamma z} \tag{1.44}$$

and

$$\underline{H}_x(z) = \frac{j\gamma}{\omega\mu}(\underline{E}_{y,h} \cdot e^{-\gamma z} - \underline{E}_{y,r} \cdot e^{\gamma z}) = \underline{H}_{x,h} \cdot e^{-\gamma z} + \underline{H}_{x,r} \cdot e^{\gamma z} \,. \tag{1.45}$$

Equation (1.33) is valid for the propagation constant, and Eq. (1.36) is valid for the characteristic wave impedance, where, because of the sign reversal in the equation for $\underline{H}_x$, is defined as

$$\underline{Z}_F = -\frac{\underline{E}_{y,h}}{\underline{H}_{x,h}} = \frac{\underline{E}_{y,r}}{\underline{H}_{x,r}} \,. \tag{1.46}$$

In the presence of both field intensity combinations $\underline{E}_x, \underline{H}_y$ and $\underline{E}_y, \underline{H}_x$, overlapping both plane waves again yields a plane wave. If both waves oscillate (assuming that their directions of polarization are perpendicular to each other, and their frequency and direction of propagation are equal, inphase or out of phase), then the individual components $\underline{E}_{x,h}$ and $\underline{E}_{y,h}$, and $\underline{H}_{y,h}$ and $\underline{H}_{x,h}$, respectively, can be geometrically added to the resulting field intensities $\vec{\underline{E}}_h$ and $\vec{\underline{H}}_h$. The resulting wave is also linearly polarized, whereby $\vec{\underline{E}}_h$ and $\vec{\underline{H}}_h$ oscillate in a fixed direction given by the ratio of wave amplitudes and are perpendicular to each other.

*$v_g = d\omega/d\beta$ is the signal velocity when the signal frequency band is so small that $\beta$ can be approached through a linear function in the desired frequency range. Otherwise, signal distortions occur and no clearly definable signal velocity can be determined.

If both waves possess a random phase shift in time with respect to each other, then the instantaneous values of the field components concerned must be combined with the resulting field intensities at the same points in time. The resulting vectors of the electrical and magnetic field intensities now describe ellipses. We will examine the special case of a phase shift of $\Delta\varphi = \pm\pi/2$. Corresponding to $E_x = E_x \cdot \sin(\omega t - \beta z)$ and $E_y = \pm E_y \cdot \cos(\omega t - \beta z)$, at zero passage, one field intensity component reaches either the maximum or the respective minimum value of the other component. Therefore, the vector of the resulting electrical field intensities describes an ellipse, and upon three-dimensional consideration an elliptical helical line is revealed. We designate this wave to be *elliptically polarized.* At $\Delta\varphi = \pi/2$ the wave runs clockwise with reference to the direction of propagation, and at $\Delta\varphi = -\pi/2$ it runs counterclockwise. If both partial waves possess the same amplitude, then the polarization ellipse turns into a circle and we speak of *circular polarization.* For $\kappa = 0$ the vector of the resulting magnetic field intensity is perpendicular to the resulting electrical field intensity vector and oscillates with it in phase. For elliptical polarization and $\kappa \neq 0$, $\vec{E}$ and $\vec{H}$ are no longer spatially perpendicular to each other.

The direction in which energy is transported for a propagating electromagnetic wave is perpendicular to the vectors of the electrical and magnetic field intensities, and is given by the direction of the vector

$$\vec{P}' = \vec{E} \times \vec{H}. \tag{1.47}$$

The vector $\vec{P}$ (in order to distinguish it from the current density, it is not designated $S$ as is usually the case) is called the Poynting vector, and it describes the power flow of the wave per unit area. According to the Poynting law, the power $\oint_{(A)} (\vec{E} \times \vec{H}) \cdot d\vec{A}$, eliminated from a volume $V$ with the surface $A$, is decreased by an amount equal to the added power $P_{zu}$ for the power used and the change in time of the accumulated electrical and magnetic energy in $V$

$$\oint_{(A)} (\vec{E} \times \vec{H}) \cdot d\vec{A} = P_{zu} - \int_{(V)} \varkappa \cdot \vec{E}^2 \cdot dV - \frac{\partial}{\partial t} \int_{(V)} (w_e + w_m) \cdot dV. \tag{1.48}$$

The complex representation of the harmonic time dependence of the field is

$$\underline{\vec{P}}' = \frac{1}{2} \underline{\vec{E}} \times \underline{\vec{H}}^* . \tag{1.49}$$

Here $\underline{\vec{H}}^*$ is the conjugated complex value of $\underline{\vec{H}}$.
We obtain for the mean time of the power flow

$$\overline{\vec{P}'} = \mathrm{Re}\,\{\underline{\vec{P}}\} = \frac{1}{2}\,\mathrm{Re}\,\{\underline{\vec{E}} \times \underline{\vec{H}}^*\} . \tag{1.50}$$

The power transported in the $z$-direction follows from the plane wave propagating in the $z$-direction which was examined above; and, for the mean time of the power transported per unit area, the following is valid (independent of $x$ and $y$)

$$\overline{P'_z} = \frac{1}{2}\,\mathrm{Re}\,\{\underline{E}_x \underline{H}_y^*\} = \frac{\hat{E}_x^2}{2 Z_F} \tag{1.51}$$

## 1.4 BOUNDARY OR LIMIT CONDITIONS, SKIN EFFECT

The electromagnetic plane wave which propagates through uniform space of infinite length is regarded as the ideal case. In practice, the wave is altered by local changes of the propagation medium, such as, for example, one of the transmission lines which serves wave propagation. At the contact surface between two media, the field must fulfill certain boundary or limit conditions. The integration constants which result from solving the field equations can be determined with the help of these conditions.

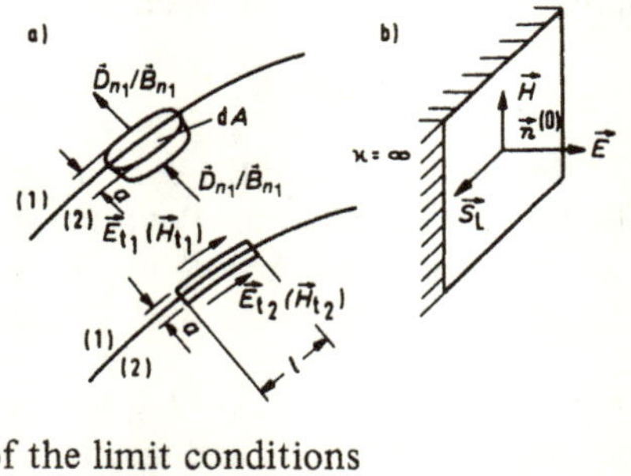


**Fig. 1.5** Illustration of the limit conditions

For deriving the limit conditions, we use the appropriate integral form of the field equations. To this end, we observe the contact surfaces between two media, 1 and 2, with the material constants $\epsilon_1, \mu_1, \kappa_1$ and $\epsilon_2, \mu_2, \kappa_2$. We stipulate first of all $\kappa_1 = \kappa_2 = 0$. If we calculate the integral $\int \vec{D} \cdot d\vec{A} = Q$ over the surface of a disk-shaped region of the contact surface, according to Fig. 1.5a, whereby the density of the disk tends from $a \to 0$ and thereby the electrical flux disappears through the narrow side, then we obtain

$$D_{n_2} \cdot dA - D_{n_1} \cdot dA = q_A \cdot dA \tag{1.52}$$

and thereby

$$D_{n_2} - D_{n_1} = q_A \tag{1.53a}$$

or, respectively,

$$\varepsilon_{r_2} \cdot E_{n_2} - \varepsilon_{r_1} \cdot E_{n_1} = \frac{q_A}{\varepsilon_0} . \tag{1.53b}$$

The difference in the normal components of the electrical shift density $\vec{D}$ on the contact surface is equal to the surface density of charge $q_A$ (charge per surface) at the interface.

Through an analogous examination of the magnetic flux $\int \vec{B} \cdot d\vec{A} = \varphi$, because of the fact that the flux coming into a boundary envelope is the same as the flux going out, we obtain the following

$$B_{n_2} - B_{n_1} = 0 \tag{1.54}$$

and, respectively,

$$\mu_{r_2} \cdot H_{n_2} - \mu_{r_1} \cdot H_{n_1} = 0 . \tag{1.55}$$

Because the magnetic field is solenoidal, the induction lines are fully self-contained, and therefore the normal component of the magnetic induction $\vec{B}$ holds steady at the contact surface.

To investigate the behavior of the tangential components of the fields at the contact surface, the laws of induction and circulation are used on the small rectangle in the boundary layer according to Fig. 1.5a. The sides of the rectangle of length $l$ should lie in either of the two media parallel to the separation line. The widths are chosen from $a \to 0$, so that the parts of the peripheral integral in question can be disregarded. The law of induction [Eqs. (1.4, (1.5)] yields

$$(E_{t_2} - E_{t_1}) \cdot l = -j\omega \cdot B \cdot l \cdot a . \tag{1.56}$$

With $a \to 0$ we obtain

$$E_{t_1} = E_{t_2} \tag{1.57}$$

or, respectively,

$$\frac{D_{t_1}}{\varepsilon_{r_1}} = \frac{D_{t_2}}{\varepsilon_{r_2}} . \tag{1.58}$$

Application of the law of circulation [Eq. (1.3)] for the small rectangular surface yields

$$(H_{t_2} - H_{t_1}) \cdot l = \left(S_L + \frac{\partial D}{\partial t}\right) \cdot l \cdot a . \tag{1.59}$$

For $a \to 0$, the second part disappears to the right side of Eq. (1.59), which is yielded by the displacement current.

If the media bordering each other are non-conductive or poorly conductive, then the first term, given by the conducting current, is zero and we obtain

$$H_{t_1} = H_{t_2} \tag{1.60}$$

At present, we shall examine the relationships of a highly conductive contact surface. In the case of alternating-current conductors, the magnetic field creates a flow of current in the conductor's interior as a result of the voltage induced at that point, which weakens the original current in the inner region and then strengthens as it approaches the surface. This effect of decreasing the current density from the conductor surface toward the interior becomes increasingly noticeable with increasing frequency, and is designated as the *current displacement,* the *Kelvin,* or the *skin effect.*

We simply assume that the magnetic field intensity and the current density decrease exponentially from the conductor surface toward the interior (Fig. 1.6).

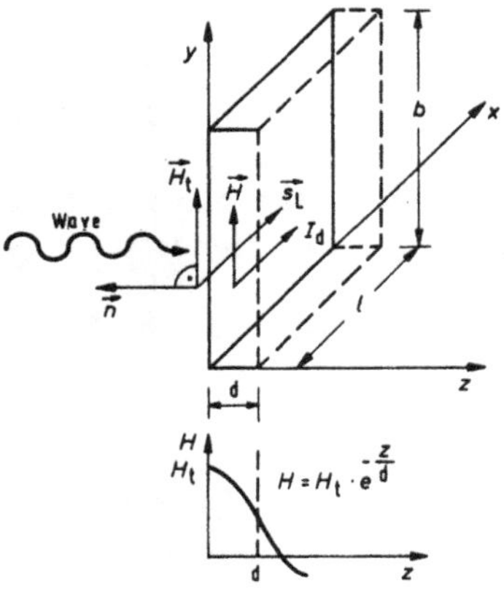


**Fig. 1.6** Illustration of the skin effect

The depth within which these strengths are decreased to 1/e of the maximum value on the conductor surface is designated as the *penetration depth d.* For the depth of penetration, which decreases with increasing frequency $f$, conductivity $\kappa$ and permeability $\mu$, the following relationship is valid:

$$d = \frac{1}{\sqrt{\pi f \mu \kappa}} \, . \tag{1.61}$$

[Compare this derivation with the section on surface waves, Eq. (1.275)]

If the contact surface is formed by a highly conductive medium 1, then the high frequency conductor current only flows into a thin layer near the surface. In this case the product of the high current density $S_L$ and the small length $a$ in Eq. (1.59) remains finite, and yields the surface current distribution $S'_L$

$$S'_L = \lim_{a \to 0} S_L \cdot a \,. \tag{1.62}$$

With this, we obtain the limit condition

$$H_{t_2} - H_{t_1} = S'_L \tag{1.63}$$

and, respectively,

$$\frac{B_{t_1}}{\mu_{r_1}} - \frac{B_{t_2}}{\mu_{r_2}} = \mu_0 \cdot S'_L \,. \tag{1.64}$$

Since the electromagnetic field disappears in the interior of an ideal conductor ($\kappa \to \infty$), we must set $H_{t_1} = 0$ in this case, and we obtain in vector form (compare with Fig. 1.5b)

$$\vec{n}^{(0)} \times \vec{H} = \vec{S}'_L \,. \tag{1.65}$$

The surface current density $S'_L$ at a point on the conductor surface is equal to the magnetic field intensity which dominates there and is tangentially directed.

A surface density of charge $q_A$ is connected to the surface current density. The shift density lines end perpendicular to this surface layer of charge. When $D_{n_1} = 0$, Eq. (1.53) for the ideal conductivity surface yields

$$D_n = q_A \tag{1.66}$$

Because of the field mobility in the interior of an ideal conductor ($\kappa \to \infty$), with $E_{t_2}, B_{n_2} = 0$ at its surface, we obtain the boundary conditions

$$E_t, D_t = 0 \tag{1.67}$$

and

$$H_n, B_n = 0 \tag{1.68}$$

(compare with Fig. 1.5b).

The condition $E_t = 0$ means that no voltage drop can occur on an infinitely conductive, current-carrying conductor.

Because of the extremely negligible penetration depth of the fields and currents for metals in the microwave domain, we can often assume infinite conductivity to be a reasonable approximation.

The influence of high but finite conductivity can be considered approximately by surface resistance, such as, for example, a calculation of the attenuation through wall current losses.

Here we simply assume that the entire wall current $I_L = S'_L \cdot b$ (Fig. 1.6), with a constant current density equal to the maximum value on the conductor surface, would flow within one layer of the conductor surface. For the thickness of this layer, we obtain the same value as for the penetration depth $d$ [Eq. (1.61)], which we therefore designate as the equivalent index bed thickness. The surface resistance amounts, therefore, to $R_d = b/\kappa \cdot 1 \cdot d$ and the related surface resistance of a surface element, whose length 1 in the direction of the current is selected to be as large as the width $b$ of the current path.

$$R_{\square} = \frac{1}{\varkappa d} = \sqrt{\frac{\pi f \mu}{\varkappa}} \qquad (1.69)$$

An equally large inductive component is yielded so that for the surface impedance we obtain

$$\underline{Z}_d = \frac{1 + j}{\varkappa \cdot d} . \qquad (1.70)$$

The following connection is valid

$$\underline{\vec{E}}_t = \underline{Z}_d \cdot \vec{S}'_L = \underline{Z}_d \cdot (\vec{n}^{(0)} \times \vec{H}) . \qquad (1.71)$$

We shall also examine the case of finite, non-zero conductances $\varkappa_1$ and $\varkappa_2$ of the media 1 and 2. In addition to Eq. (1.53), the following must also hold true [compare with Eq. (1.19)]:

$$\varkappa_2 \cdot \underline{E}_{n_2} - \varkappa_1 \cdot \underline{E}_{n_1} = -j\omega q_A . \qquad (1.72)$$

With Eq. (1.53), this yields

$$\underline{E}_{n_1} \cdot (\varkappa_1 + j\omega\varepsilon_1) - \underline{E}_{n_2}(\varkappa_2 + j\omega\varepsilon_2) = 0 . \qquad (1.73)$$

## 1.5 TEM-MODES (L-WAVES)

### 1.5.1 Field Examination

If we were to introduce very thin, ideally conductive, metal plates of infinite length perpendicular to the electrical energy lines in an electromagnetic plane wave field, the field distribution would be maintained because the limit conditions at the metal plates $E_t, H_n = 0$ would be satisfied. Consequently, an electromagnetic wave can propagate between two unlimited, highly conductive parallel metal plates with the field distribution in the $z$-direction, as outlined in Fig. 1.7. The field distribution can be understood as the segment of a plane wave. The wave is directed by this *two-wire circuit*.

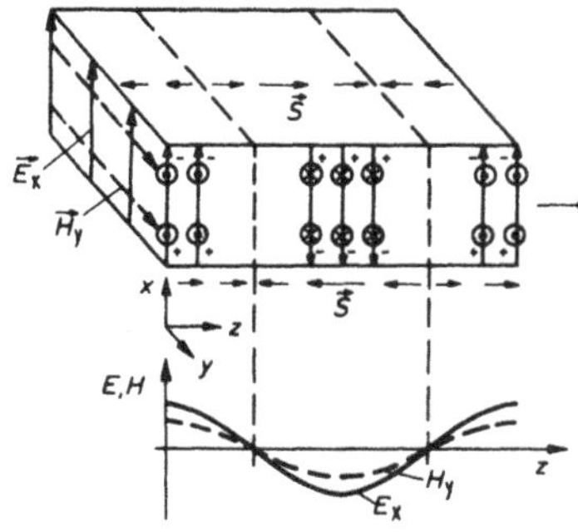


**Fig. 1.7** Two-wire circuit with field charge, and surface current density distributions

In the assumed theoretical borderline case when $\kappa \to \infty$, the conductor interior is field-free, and the electrical and magnetic fields possess only components diagonal to the direction of propagation. Thus, we obtain a *transverse electromagnetic wave* (TEM-wave).

Waves that propagate along practical circuits and for which the longitudinal field components $E_z, H_z \to 0$ disappear in the borderline case when $\kappa \to \infty$ are called *TEM-modes* or *L-waves,* also known as *Lecher waves* (from E. Lecher). The surface currents flowing along the inner wire surfaces provide transport for the surface charges located there.

If the length of wires $a$ is smaller than half the free space wavelength, $a < \lambda_0/2$, then this TEM-wave is the only wave capable of propagating through the two-wire medium. In the case where the wire length increases – or, rather, a reduction of $\lambda_0$, which corresponds to an increase in the wave frequency – more and more wavemodes (hollow waveguide modes) become capable of propagating, as shall be explained in sec. 1.6.

By limiting the theoretical two-wire circuit to the two parallel conductor lines, the practical *stripline* is created (Fig. 1.8). The field distribution in the planes perpendicular to the direction of propagation is now inhomogeneous, as in the case of TEM-waveguides described below. In the case of the assumed ideal conductor, the electrical and magnetic fields remain transverse to the direction of propagation.

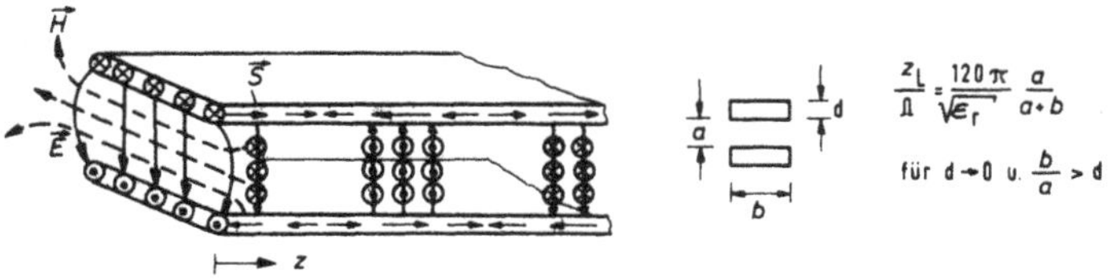


**Fig. 1.8** Stripline

The charges and conduction current are principally concentrated on the inner conductor surfaces.

Circuits which present constant dimensions perpendicular to the wave's direction of propagation and a uniform dielectric are designated as *homogeneous* circuits. Microstrip lines created from the striplines will be dealt with in a separate section because of their special technical significance.

In Fig. 1.9 the cross section of another homogeneous waveguide is represented, the *balanced feeder* (Lecher system) with the field distribution of a TEM-wave.

$$\frac{z_L}{\Omega} = \frac{120}{\sqrt{\varepsilon_r}} \; \text{arcos} \; \frac{a}{D} =$$

$$= \frac{120}{\sqrt{\varepsilon_r}} \; \ln \left[ \frac{a}{D} + \sqrt{\left(\frac{a}{D}\right)^2 - 1} \; \right]$$

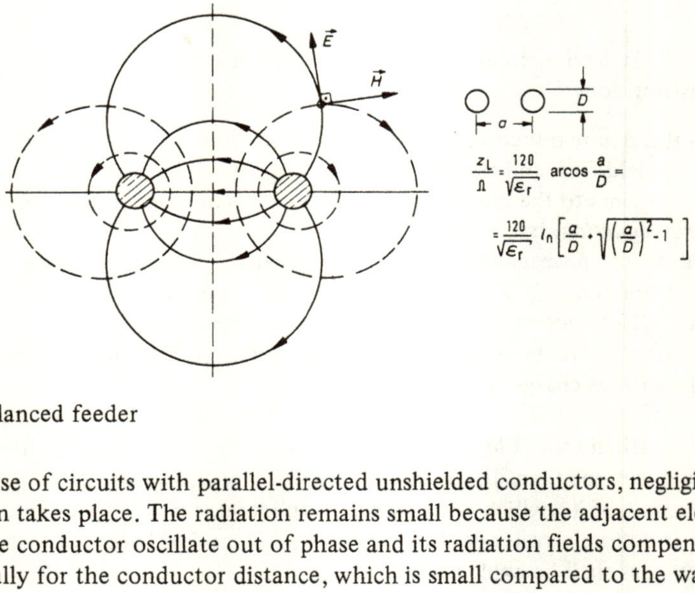

**Fig. 1.9** Balanced feeder

In the case of circuits with parallel-directed unshielded conductors, negligible radiation takes place. The radiation remains small because the adjacent elements of the conductor oscillate out of phase and its radiation fields compensate externally for the conductor distance, which is small compared to the wavelength.

A waveguide often used in practice is the *coaxial transmission line,* for which the field distribution of a TEM-wave is shown in Fig. 1.10. We may think of a coaxial transmission line as having arisen out of the coiling together of the stripline. It offers a good shielding effect. For circuit representation, formulas giving the waveguide resistances [see Eqs. (1.80) and (1.81)] for lossless circuits are specified.

The carrier of the energy transported horizontally along a circuit is the electromagnetic field. It is the conductor's function to channel this energy in the desired direction.

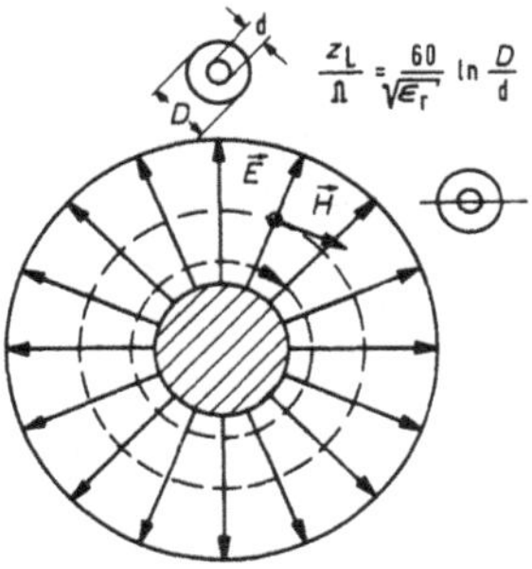


**Fig. 1.10** Coaxial transmission line

In the assumed case of conductivity and homogeneous, unstratified dielectric, the interior of the conductor is field-free, the electrical and magnetic fields are transverse, and the Poynting vector $\vec{P}' = \vec{E} \times \vec{H}$ of the equalizing current possesses only one component in the wave's direction of propagation and is zero within the conductor.

At high frequencies the magnetic field within the conductor's interior is virtually zero; the magnetic energy lines of the circuit operate as balanced lines.

In practice, the conductor material possesses a *finite conductivity* $\kappa_L$. Then the high frequency electromagnetic field penetrates the metal somewhat. The current flowing in the $z$-direction is therefore connected with a horizontal component $E_z$ of the electrical field intensity because $\vec{S} = \kappa \cdot E$. With the exception of the coaxial circuit with its cylindrically symmetrical field, the electrical field also receives a tangential component and the magnetic field a normal component. Hence, the conductor contour is no longer a balanced line and the magnetic energy lines no longer agree with the balanced lines of the steady-state field. In addition to the horizontal currents in the conductor, we also find reactive currents which are a continuation of the shift currents in dielectric for the flux. With two-wire circuits, the current density is raised in the adjacent ranges (proximity effect). Finally, except for the case of the coaxial transmission line, a horizontal magnetic field intensity $H_z$ also appears, which is coupled with the transverse shift and conductor currents, leading to eddy-current losses. Consequently, in the general case of finite conductivity of the conductor material, six field components in all are present, and the Poynting vector of the equalizing current possesses components in all three directions in space.

There is, therefore, no pure TEM-wave propagation. However, this can be considered as a very good approximate solution.

With an assumed pure TEM-wave propagation the conductor interior is field-free, and the electrical and magnetic energy lines run only in planes which are perpendicular to the direction of propagation. In such a plane, through which no electrical and magnetic field can cross, an electrical voltage $U_{AB}$ between both conductors and the current $I$, which flows in a conductor through this plane, can be clearly defined. In determining these strengths through integration in the plane of cross section, the integration between points A and B of the conductor for the determination of the voltage

$$U_{AB} = \int_A^B \vec{E} \cdot d\vec{s}$$

and the contour integral at one of the conductors for the determination of the conductor current $I = \oint \vec{H} \cdot d\vec{s}$ are independent of the special path direction. In the case of finite conductivity, this unambiguous solution is lost: the current integral is a function of the integration path because potential differences occur in the conductor cross section, and as a result of the current induced by $H_z$. The current integral becomes dependent on the path because of the horizontal shift currents that occur there.

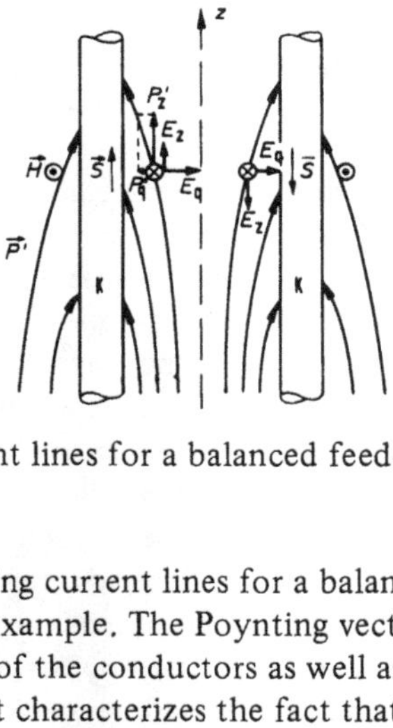


**Fig. 1.11** Equalizing current lines for a balanced feeder with finite conductivity of the conductor

In Fig. 1.11 the equalizing current lines for a balanced feeder in a horizontal section are outlined as an example. The Poynting vector receives a component which lies in the direction of the conductors as well as the component in the direction of propagation. It characterizes the fact that a portion of the energy in the conductor directed by the wave is converted into heat. In this connection the current density of the energy in the direction of wave propagation decreases and the wave becomes attenuated. The component $P'_z$ of the Poynting vector in the direction of propagation is formed from the transverse components $E_q$ and $H$, and the component $P'_q$ from $E_z$ and $H$.

The waveguides dealt with here then direct only L-waves if the sizes of their cross-sections are small compared to the wavelength. Those wave modes possible at higher frequencies will be dealt with in sec. 1.6 on E- and H-waves in hollow waveguides.

### 1.5.2 Line Theory

TEM-modes can be easily and precisely described with the transmission equations known from line theory with the help of voltage and current waves, when for high $\kappa_L$ ($\kappa_L \gg \omega_\epsilon$) cross-section dimensions of the circuit, which are small in comparison to the wavelength, the dielectric losses resulting from $E_z$ and line loss, and the dielectric losses in the conductor are negligible. In this manner, the line losses and dielectric losses in the conductor are also created by the electrical cross field, as well as the losses in the dielectric and the conductor as a result of the transverse electrical alternating fields induced by $H_z$.

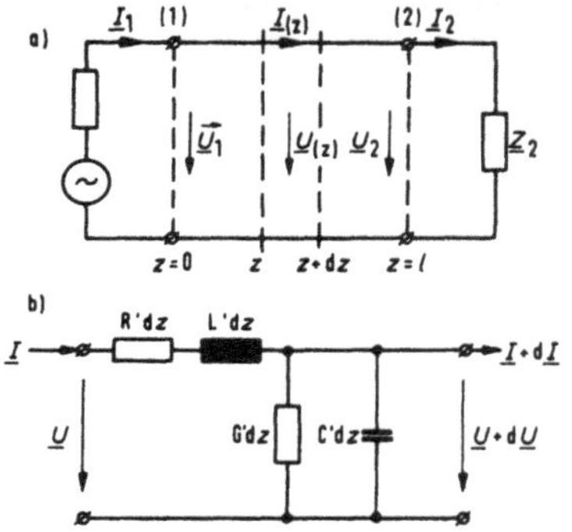


**Fig. 1.12** L-waveguide with equivalent circuit for a differential cross section

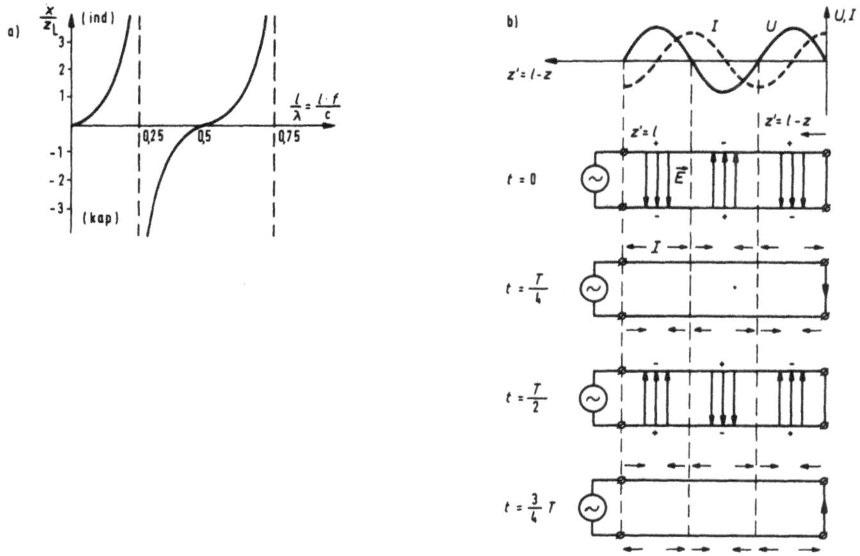


**Fig. 1.13** Short circuit: (a) input resistance, (b) current, voltage, and electrical field strength distribution

Figure 1.12 represents an equivalent circuit for a cross section with the differential length dz of an L-waveguide (for example, a coaxial transmission line or a balanced feeder). The circuit is homogeneous, which means that the geometry and material characteristics are the same in every plane of cross section, or the propagation constant is independent of the position $z$. The propagation constants are dependent on frequency. The resistance $R'$, which takes into consideration the ohmic losses in the conductors, increases as a result of the current shift, and does so at high frequencies approximately proportional to $\sqrt{f}$. The conductance per unit length $G'$, which represents the dielectric losses, increases approximately proportional to $f$, and even more strongly at high frequencies. The inductance $L'$ and the distributed capacity $C'$ generally change only slightly with frequency, $C'$ decreases as a result of the dielectric losses and $L'$ decreases because of the drop in inner conductivity.

Kirchhoff's equations yield the following relations according to Fig. 1.13:

$$\frac{\partial \underline{U}}{\partial z} = -\left(R'\underline{I} + L'\frac{\partial \underline{I}}{\partial t}\right) \tag{1.74}$$

and

$$\frac{\partial \underline{I}}{\partial z} = -\left(G'\underline{U} + C\frac{\partial \underline{U}}{\partial t}\right), \tag{1.75}$$

with which the differential equation (telegraphic equation) for a harmonic time dependency with a cyclic frequency $\omega$

$$\frac{\partial^2 \underline{U}}{\partial z^2} - \gamma^2 \underline{U} = 0 \tag{1.76}$$

can be established.

Here, for the so-called *propagation constant* (transmission constant), the following holds:

$$\gamma = \alpha + \mathrm{j}\beta = \sqrt{(R' + \mathrm{j}\omega L')(G' + \mathrm{j}\omega C')}\,. \tag{1.77}$$

The differential equations (1.74) and (1.75), respectively, have the same form as Eq. (1.29) and the solution

$$\underline{U}(z) = \underline{U}_{1\mathrm{h}} \cdot \mathrm{e}^{-\gamma z} + \underline{U}_{1\mathrm{r}}\mathrm{e}^{\gamma z} = \underline{U}_{\mathrm{h}}(z) + \underline{U}_{\mathrm{r}}(z)\,. \tag{1.78}$$

For the current $\underline{I}$ we obtain

$$\underline{I}(z) = \frac{\underline{U}_{1\mathrm{h}}}{\underline{Z}_{\mathrm{L}}}\mathrm{e}^{-\gamma z} - \frac{\underline{U}_{1\mathrm{r}}}{\underline{Z}_{\mathrm{L}}}\mathrm{e}^{\gamma z} = \frac{\underline{U}_{\mathrm{h}}(z)}{\underline{Z}_{\mathrm{L}}} - \frac{\underline{U}_{\mathrm{r}}(z)}{\underline{Z}_{\mathrm{L}}} = \tag{1.79}$$

$$= \underline{I}_{1\mathrm{h}}\mathrm{e}^{-\gamma z} + \underline{I}_{1\mathrm{r}} \cdot \mathrm{e}^{\gamma z} = \underline{I}_{\mathrm{h}}(z) + \underline{I}_{\mathrm{r}}(z)\,.$$

Now, therefore, there is a superposition of two advancing waves for the voltage variation and the current path: a wave propagating in the $z$-direction is designated as an advancing wave (index h), or incident wave, and a wave propagating in the opposite of the $z$-direction is designated as a receding or reflected wave.

In the case of circuits with a clearly defined L-wave, the relationship of the complex amplitudes of the voltage and the current in a cross section of the conducting wire is defined as the *line resistance* $\underline{Z}$. If, with these amplitudes, we are dealing with those of an incident wave, then we designate the line resistance as the *characteristic line impedance* $\underline{Z}_L$. The following is then valid [according to Eq. (1.79)]:

$$\underline{Z}_L = \frac{\underline{U}_{1h}}{\underline{I}_{1h}} = -\frac{\underline{U}_{1r}}{\underline{I}_{1r}} . \tag{1.80}$$

For the L-wave under consideration, we obtain

$$\underline{Z}_L = \sqrt{\frac{R' + j\omega L'}{G' + j\omega C'}} . \tag{1.81}$$

In the case of the complex propagation constant $\gamma = \alpha + j\beta$, we designate the magnitude $\alpha$, which specifies the change in amplitude of the wave per unit of length, as the *attenuation constant*, and the magnitude $\beta$, which represents the phase transition per unit of length, as the *phase constant.*

In the case of a circuit of length $l$ with $\gamma l = \alpha l + j\beta l = a + jb$, we designate $\gamma l$ as the transmission equivalent, $a$ as the attenuation equivalent or transmission attenuation, and $b$ as the impedance angle or the transmission angle of the circuit.

For lossless circuits ($R' = G' = 0$) and, approximately, for low-loss circuits or at high frequencies, the following hold true:

$$\beta = \omega \cdot \sqrt{L'C'} \tag{1.82}$$

and

$$Z_L = \sqrt{\frac{L'}{C'}} . \tag{1.83}$$

For negligible losses, the attenuation constant can be approximately calculated according to

$$\alpha = \alpha_R + \alpha_G = \frac{1}{2} \cdot \frac{R'}{Z_L} + \frac{1}{2} G' \cdot Z_L . \tag{1.84}$$

The attenuation $\alpha_R$ resulting from ohmic losses is designated as the resistance loss, the current attenuation, or the horizontal damping. The attenuation $\alpha_G$ resulting from the dielectric losses is designated as leakage, the voltage attenuation, or the transverse attenuation.

For small dielectric losses ($\tan \delta_\epsilon \ll 1$) $\alpha_G$ can be calculated with

$$\alpha_G = \frac{1}{2} \beta \cdot \tan \delta_\varepsilon = \frac{1}{2} \omega \cdot \sqrt{L'C'} \cdot \tan \delta_\varepsilon \, . \tag{1.85}$$

$\alpha_G$ is approximately proportional to the frequency $f$, and $\alpha_R$, as a result of the skin effect, is approximately proportional to $\sqrt{f}$.

The time dependence of the circuit's voltage and current is not expressed in Eqs. (1.78) and (1.79). It can be described through multiplication with the complex time factor $e^{j\omega t}$. In real notation, for example, the following results for the incident wave according to the formation of the real part ($\underline{U}_{1h}$ being real):

$$U_h(z, t) = \mathrm{Re}\,\{\underline{U}_{1h} \cdot e^{-\gamma z} \cdot e^{j\omega t}\} = \hat{u}_{1h} \cdot e^{-\alpha z} \cdot \cos(\omega t - \beta z) \, . \tag{1.86}$$

The shortest distance between two points with the same phase relationship on a wave propagating along the line is designated as the line wavelength $\lambda_L$.

With $\beta \cdot \lambda = 2\pi$ we obtain

$$\lambda_L = \frac{2\pi}{\beta} \, . \tag{1.87}$$

The speed with which a certain phase condition propagates along the line, the so-called *phase velocity* $v_p$, with $\omega t - \beta z = \text{const}$, yields upon differentiation $\omega dt - \beta dz$ and with Eq. (1.82) for the low-loss circuit

$$v_p = \frac{dz}{dt} = \frac{\omega}{\beta} = \frac{1}{\sqrt{L'C'}} \, . \tag{1.88}$$

For the *group velocity* $v_g$ at which the energy, or a signal which comes from a small frequency group, propagates along the line, we obtain the same value as with the low-loss L-waveguide

$$v_g = \frac{d\omega}{d\beta} = \frac{1}{\sqrt{L'C'}} \, . \tag{1.89}$$

With the material constants $\epsilon = \epsilon_r \cdot \epsilon_0$ and $\mu = \mu_r \cdot \epsilon_0$ of the circuit's dielectric, $L'C' = \mu\epsilon$ is yielded, and we obtain (1.90).

$$\lambda_L = \frac{\lambda_0}{\sqrt{\mu_r \cdot \varepsilon_r}} \, , \tag{1.90}$$

$$v_p = v_g = \frac{c}{\sqrt{\mu_r \cdot \varepsilon_r}} = c \cdot \frac{\lambda_L}{\lambda_0} = f \cdot \lambda_L \tag{1.91}$$

$$Z_L = \frac{1}{v_p \cdot C'} = v_p \cdot L' \tag{1.92}$$

and

$$\alpha_G = \pi f \cdot \sqrt{\mu_r \varepsilon_r} \cdot \tan \delta_\varepsilon \,. \tag{1.93}$$

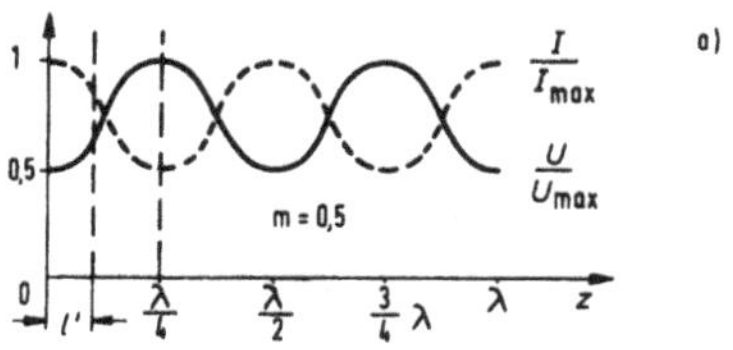


**Fig. 1.14a** Voltage distribution, current distribution, and random phase along a line with a real terminal resistance [see also Figs. 1.14(b) and (c), Ch. 1, sec. 1.5, between Eqs. (1.124) and (1.125)]

Let us consider a circuit (or the cross section of a circuit) with the characteristic line impedance $\underline{Z}_L$ and of length $z = l$. The output gate is loaded with the terminal resistance $\underline{Z}_2$ (Fig. 1.14). At present we will investigate the voltage and current distributions along the line and the impedance at the input of the circuit (or the circuit cross section under consideration) at gate (1). At gates (1) and (2), respectively, we obtain from Eqs. (1.78) and (1.79)

$$z = 0: \quad \underline{U}_1 = \underline{U}_{1h} + \underline{U}_{1r} \tag{1.94}$$

$$\underline{I}_1 = \frac{\underline{U}_{1h}}{\underline{Z}_L} - \frac{\underline{U}_{1r}}{\underline{Z}_L} \tag{1.95}$$

$$z = l: \quad \underline{U}_2 = \underline{U}_{1h} \cdot e^{-\gamma l} + \underline{U}_{1r} \cdot e^{\gamma l} = \underline{U}_{2h} + \underline{U}_{2r} \tag{1.96}$$

$$\underline{I}_2 = \frac{\underline{U}_{1h}}{\underline{Z}_L} \cdot e^{-\gamma l} - \frac{\underline{U}_{1r}}{\underline{Z}_L} e^{\gamma l} = \underline{I}_{2h} + \underline{I}_{2r} = \frac{\underline{U}_{2h}}{\underline{Z}_L} - \frac{\underline{U}_{2r}}{\underline{Z}_L} \tag{1.97}$$

After eliminating $\underline{U}_{1h}$ and $\underline{U}_{1r}$ and introducing the hyperbolic functions, we obtain the *transmission equations* for the section of a line of length $l$ and characteristic line impedance $\underline{Z}_L$ in the chain form

$$\underline{U}_1 = \underline{U}_2 \cdot \cosh \gamma l + \underline{I}_2 \cdot \underline{Z}_L \cdot \sinh \gamma l \,, \tag{1.98}$$

$$\underline{I}_1 = \underline{I}_2 \cdot \cosh \gamma l + \frac{\underline{U}_2}{\underline{Z}_L} \cdot \sinh \gamma l \,. \tag{1.99}$$

The relationship between the complex tension or the electrical cross-field intensity of the reflected wave and the complex tension or electrical cross-field intensity of the wave propagating toward the receiver in a certain cross section of the conducting wire is defined as the *reflectivity* $r$. It is

$$\underline{r}(z) = \frac{\underline{U}_r(z)}{\underline{U}_h(z)} = \frac{\underline{E}_{q,r}(z)}{\underline{E}_{q,h}(z)} = -\frac{\underline{I}_r(z)}{\underline{I}_h(z)} = -\frac{\underline{H}_{q,r}(z)}{\underline{H}_{q,h}(z)} \,. \tag{1.100}$$

For the corresponding power ratio, the following holds true

$$\frac{P_r}{P_h} = \left| \frac{\underline{U}_r}{\underline{U}_h} \right|^2 = r^2 \,. \tag{1.101}$$

For the relation between the reflectivity $\underline{r}(\underline{Z})$ and the line resistance $\underline{Z}(z)$ at point $z$, we obtain with Eqs. (1.78) and (1.79)

$$\underline{Z}(z) = \frac{\underline{U}(z)}{\underline{I}(z)} = \underline{Z}_L \frac{\underline{U}_h(z) + \underline{U}_r(z)}{\underline{U}_h(z) - \underline{U}_r(z)} = \underline{Z}_L \frac{1 + \dfrac{\underline{U}_r(z)}{\underline{U}_h(z)}}{1 - \dfrac{\underline{U}_r(z)}{\underline{U}_h(z)}} = \underline{Z}_L \frac{1 + \underline{r}(z)}{1 - \underline{r}(z)} \tag{1.102}$$

and, respectively,

$$\underline{r} = r \cdot e^{\varphi} = \frac{\underline{Z} - \underline{Z}_L}{\underline{Z} + \underline{Z}_L} \,. \tag{1.103}$$

If we use the field resistance for $\underline{Z}$ and the characteristic wave impedance $\underline{Z}_F$ for the characteristic line impedance $\underline{Z}_L$, then we obtain the same relation between $\underline{Z}$ and $\underline{r}$.

For the closed terminal circuit $z = l$ with the resistance $\underline{Z}_2$, the following holds true:

$$\underline{U}_2 = \underline{I}_2 \cdot \underline{Z}_2 \,. \tag{1.104}$$

Thereby we obtain for the reflectivity of the transmission line

$$\underline{r}_2 = \frac{\underline{Z}_2 - \underline{Z}_L}{\underline{Z}_2 + \underline{Z}_L} \,. \tag{1.105}$$

The reflected wave disappears for

$$\underline{r}_2 = 0 \,, \quad \text{that means} \quad \underline{Z}_2 = \underline{Z}_L \,. \tag{1.106}$$

In the case of a *wave adapter,* in which the circuit and its oscillation resistance are closed, the energy of the wave approaching the terminal resistance will be completely converted there.

In the case of an open transmission line, we obtain with $\underline{Z}_2 \to \infty$

$$\underline{r}_2 = 1 \quad \text{and} \quad \underline{U}_{2r} = \underline{U}_{2h} = \frac{\underline{U}_2}{2}\,; \quad \underline{I}_{2r} = -\underline{I}_{2h}\,; \quad \underline{I}_2 = 0\,. \tag{1.107}$$

In the case of a closed-circuit such that

$$\underline{r}_2 = -1 \quad \text{und} \quad \underline{U}_{2r} = -\underline{U}_{2h}, \quad \underline{U}_2 = 0; \quad \underline{I}_{2r} = \underline{I}_{2h} = \frac{\underline{I}_2}{2}\,. \tag{1.108}$$

In both cases the energy of the wave at the transmission line cannot be converted (cross-section size of the circuit $\ll \lambda$, which means reflection can be ignored) and is completely reflected.

We will now examine the transformation of the reflectivity along the length of the circuit to the generator side. If we establish a relationship between a reflected and an incident voltage wave at the origin of the line segment closed with $\underline{Z}_2$ and of length $l$, then at that point we obtain with Eq. (1.96) for the reflectivity

$$\underline{r}_1 = \frac{\underline{U}_{1r}}{\underline{U}_{1h}} = \frac{\underline{U}_{2r} \cdot e^{-\gamma l}}{\underline{U}_{2h} \cdot e^{\gamma l}} = \underline{r}_2 \cdot e^{-2\gamma l} = \underline{r}_2 \cdot e^{-2\alpha l} \cdot e^{-j2\beta l}\,. \tag{1.109}$$

With low-loss circuits where $\alpha l \approx 0$ this transformation occurs quite simply: the value of the reflectivity remains constant along the circuit and only a phase shift takes place

$$\underline{r}_1 = \underline{r}_2 \cdot e^{-j2\beta l} = \underline{r}_2 \cdot e^{-j\,4\pi/\lambda \cdot l}\,. \tag{1.110}$$

We will now follow the transformation of the terminal resistance of a circuit to the circuit's origin. For the input resistance of a line segment of length $l$ and characteristic line impedance $\underline{Z}_L$, which is closed with the resistance $\underline{Z}_2$, we obtain with the help of Eqs. (1.98), (1.99), and (1.104)

$$\underline{Z}_1 = \frac{\underline{U}_1}{\underline{I}_1} = \frac{\underline{Z}_2 \cdot \cosh \gamma l + \underline{Z}_L \cdot \sinh \gamma l}{\cosh \gamma l + \dfrac{\underline{Z}_2}{\underline{Z}_L} \cdot \sinh \gamma l} = \frac{\underline{Z}_2 + \underline{Z}_L \cdot \tanh \gamma l}{1 + \dfrac{\underline{Z}_2}{\underline{Z}_L} \cdot \tanh \gamma l}\,. \tag{1.111}$$

With high frequency circuits of short length, we can often ignore the attenuation. When $\alpha l = 0$, $\gamma l = j\beta l$ and in Eqs. (1.98), (1.99), and (1.11) we can write $\cosh \gamma l = \cos \beta l$, $\sinh \gamma l = j \sin \beta l$ and $\tanh \gamma l = j \tan \beta l$. The characteristic line impedance becomes real $Z_L = \sqrt{(L'/C')}$. By gauging the oscillation resistance and $\beta = 2\pi/\lambda$, we therefore, obtain for the low-loss circuit

$$\frac{\underline{Z}_1}{Z_L} = \frac{\frac{\underline{Z}_2}{Z_L} + j\tan\frac{2\pi}{\lambda}\cdot l}{1 + j\frac{\underline{Z}_2}{Z_L}\cdot\tan\frac{2\pi}{\lambda}\cdot l}\,. \tag{1.112}$$

Now we will examine some special cases:

a) adapter $\underline{Z}_2 = Z_L$ (or $1 \to \infty$) $\quad \underline{Z}_1 = \underline{Z}_2 = Z_L$ (1.113)

b) short-circuit transmission line
$\underline{Z}_2 = 0$: $\quad \underline{Z}_1 = jZ_L \cdot \tan\beta l$ (1.114)

c) open transmission line $\underline{Z}_2 \to \infty$ $\quad \underline{Z}_1 = -jZ_L \cdot \cot\beta l$ (1.115)

d) $l = \frac{\lambda}{4}$: $\quad \underline{Z}_1 = \frac{Z_L^2}{\underline{Z}_2}$ or $\frac{\underline{Z}_1}{Z_L} = \frac{Z_L}{\underline{Z}_2}$ (Inversion) (1.116)

e) $l = \frac{\lambda}{2}$: $\quad \underline{Z}_1 = \underline{Z}_2\,.$ (1.117)

As an example, we will more closely investigate the low-loss short-circuit case. In Fig. 1.13a the path of the input resistance is applied as a function of $l/\lambda = l \cdot f/c$. The zero positions correspond to the occurrence of series resonances and the pole positions belong to the parallel resonances.

In the calculation of the voltage and current distributions along a circuit, we proceed appropriately from the transmission line. For the low-loss short-circuit, we obtain with Eq. (1.79) $\underline{I}_{1h} = \underline{I}_{2h} \cdot e^{j\beta l}$, $\underline{I}_{1r} = \underline{I}_{2r} \cdot e^{-j\beta l}$ and with Eq. (1.108)

$$\underline{I}(z) = \frac{\underline{I}_2}{2}\left[e^{j\beta(l-z)} + e^{-j\beta(l-z)}\right] = \underline{I}_2 \cdot \cos\beta(l-z) \tag{1.118}$$

and with Eq. (1.78)

$$\underline{U}(z) = j\underline{I}_2 \cdot Z_L \cdot \sin\beta(l-z)\,. \tag{1.119}$$

The path of the voltage and the current, and thereby of the electrical and magnetic field intensities along a lossless short-circuit, is represented in Fig. 1.13b.

The time dependence can be obtained through multiplication with the complex time factor $e^{j\omega t}$. As Eqs. (1.118) and (1.119) show, the voltage has a forward phase shift of 90° with respect to the current. The energy oscillates back and forth between the electrical and magnetic fields. Figure 1.13b also shows the distribution of the electrical field intensity and the current along the short circuit for four consecutive points in time at $T/4$ intervals. As a result of the complete reflection of the wave in the transmission line and through the superposition of an incedent and a reflected wave, the picture of a *standing wave* with oscillation nodes and anti-nodes of $U$ and $I$, or $E$ and $H$, is created.

Analogous relationships are obtained for the open-ended circuit which correspond to those of the short circuit extended to an equivalent length $l' = \lambda/4$ (at the transmission line $I = 0$, $U = U_{max}$).

The diagrams depicted in Fig. 1.13 are also valid for a closed-circuit with an arbitrary reactance, considering that the reactance $jX_2$ can be replaced by an equivalent line segment which is short-circuited at the terminal end, and which has the same oscillation resistance and length $l'$, according to $X_2 = Z_L \cdot \tan 2\pi l'/\lambda$. In the case of the terminal resistance $l' < \lambda/4$ and with a closed capacitance $\lambda/4 < l' < \lambda/2$.

In the case of a circuit which is not closed by its oscillation resistance, a standing-wave ratio occurs through the superposition of an advancing wave and the receding part in the distribution of voltage and current amplitudes along the circuit.

Quotients of maximum and minimum voltage or electrical cross field intensity as well as the relationship between maximum and minimum current or magnetic cross field intensity have been defined as the *standing wave ratio s* (also VSWR for *Voltage Standing Wave Ratio*)

$$s = \frac{U_{max}}{U_{min}} = \frac{E_{max}}{E_{min}} = \frac{I_{max}}{I_{min}} = \frac{H_{max}}{H_{min}} . \tag{1.120}$$

One also designates the inverse value as the *adaptation factor m*

$$m = \frac{1}{s} . \tag{1.121}$$

For the relation to the value $r$ of reflectivity, the following holds true:

$$s = \frac{1 + r}{1 - r} \tag{1.122}$$

or

$$r = \frac{s - 1}{s + 1} = \frac{1 - m}{1 + m} . \tag{1.123}$$

In the case of the wave adaptation (no reflected wave, $\underline{Z}_2 = \underline{Z}_1$), $r = 0$, $s = m = 1$ and in the case of a complete reflection (standing wave, short circuit, idling or closed-circuit with a reactance), $r = 1$, $s \to \infty$, $m = 0$.

For the section of the input of a circuit which is not reflected, $(s - 1/s + 1)^2$ is valid, and for the section which is converted into the ballast resistor, we obtain $4s/(1 + s)^2$.

The following connection is valid for the relationship between maximum or minimum voltage and current values:

$$\frac{U_{\text{max}}}{I_{\text{max}}} = \frac{U_{\text{min}}}{I_{\text{min}}} = Z_{\text{L}} \,. \tag{1.124}$$

We will now examine the voltage and current distribution along the lossless circuit for a real terminal resistance $R_2$. For $R_2 < Z_{\text{L}}, m = R_2/Z_{\text{L}}$ and for $R_2 > Z_{\text{L}}$, $s = R_2/Z_{\text{L}}$. In Fig. 1.14 the path of the relative voltage and current values $U/U_{\text{max}}$ and $I/I_{\text{max}}$ is represented for $R_2 < Z_{\text{L}}$ ($R_2 = 0.5 Z_{\text{L}}$). A corresponding path is obtained for $R_2 > Z_{\text{L}}$, whereby the axis of ordinates shifts to the right by $l' = \lambda/4$.

The case of a complex terminal resistance can be traced back to the case when $R_2 = m \cdot Z_{\text{L}}$, through the addition of an equivalent line segment of length $l'$, whereby for the phase inductance $l' < \lambda/4$ and for the phase capacitance $\lambda/4 < l' < \lambda/2$.

In Fig. 1.14b the relative voltage variation and the phase shift along a lossless circuit are represented for various $m$-values.

In the case of an attenuated circuit, the maxima and the minima increase in the direction of the circuit's origination point, whereby the maximum (and minima) of the voltage and the current are shifted further by $\lambda/4$. Thus, $m$ is now to be defined for a specified position on the circuit as the quotient of the corresponding values on the limit (or boundary) curve which links the maxima or minima, respectively. Therefore, $m$ increases in the direction of the circuit's origin.

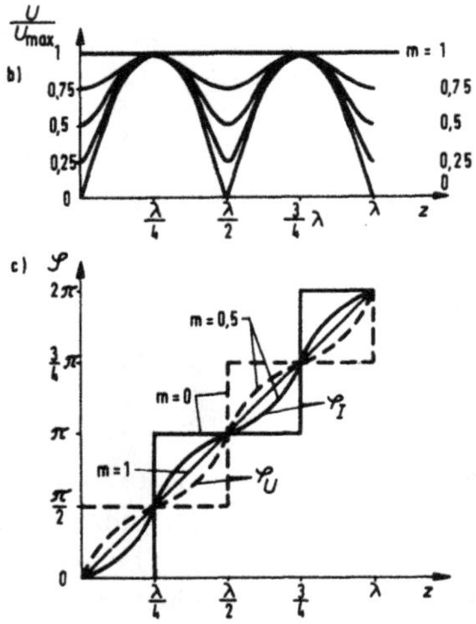


**Fig. 1.14(b) and (c)** Voltage distribution, current distribution, and random phase along a circuit for a real terminal resistance

The simple variation of the reflectivity along the lossless circuit, which only occurs by changes in angle (Eq. (1.110)), is used in circuit-circle diagrams. The reflectivity chart according to P. Smith (1939) is often used for the graphic solution of resistance transformation in circuits.

In this Smith chart the right complex resistance half-plane $R/Z_L > 0$, which applies to resistances normalized by the real characteristic impedance $\underline{Z}/Z_L = R/Z_L + j(X/Z_L)$ (or conductances) on the plane of the complex reflectivity $\underline{r} = r \cdot e^{j\varphi}$ with $r > 1$ (passive elements) is represented, therefore, in the interior of the circle of unit radius. For this representation the following holds true according to Eq. (1.102):

$$\underline{r} = \frac{\dfrac{\underline{Z}}{Z_L} - 1}{\dfrac{\underline{Z}}{Z_L} + 1} . \qquad (1.125)$$

In Fig. 1.15 illustrations of some equations in straight form $R/Z_L$, $X/Z_L$ = const are represented. Impedances with an inductive reactance $X > 0$ (and positive real part) are represented in the upper semicircle, and those with a capacitative reactance $X < 0$, in the lower semicircle. $P_A$ is the point for the adaptation $\underline{Z} = Z_L$, or $\underline{r} = 0$, $P_k$ is the short circuit point with $Z = 0$ or $\underline{r} = -1$ and $P_1$ is the no-load point with $Z \to \infty$ or $\underline{r} = 1$.

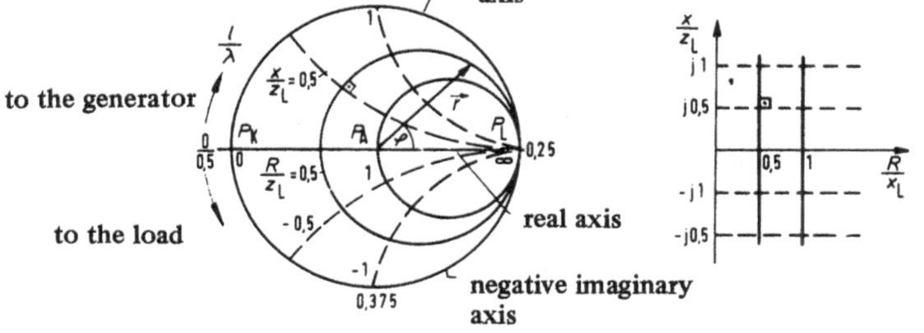


**Fig. 1.15** Smith Chart

With the use of this Smith chart for example, the input resistance of a closed line segment with the resistance $Z_2$, of length $l$, and with a characteristic line impedance $Z_L$ in the following manner: we search for the value $\underline{Z}_2/Z_L$ normalized to $Z_L$ on the chart

$$\text{reflectivity } \underline{r}_2 = \frac{\underline{Z}_2/Z_L - 1}{\underline{Z}_2/Z_L + 1} .$$

We complete the transformation to the circuit's origin by shifting the corresponding indicator at the angle $2\beta l = 4\pi l/\lambda$ in the clockwise direction ($\underline{r} = \underline{r}_2 \cdot e^{-2\beta l}$) and we read there the normalized input resistance (return transformation $\underline{Z}_1/Z_L = 1 + \underline{r}_1/1 - \underline{r}_1$).

Because of the relation $\underline{Z}_1/Z_L = Z_L/Z_2 = Y_1 \cdot Z_L$ [Eq. (1.116)] which is valid for a $\lambda/4$ circuit. We obtain the transition between resistance and conductance (inversion) in a simple manner by a change of $l/\lambda = 1/4$, or, in other words, by a reflection with respect to the central point of the chart. In the case of lossy circuits, the indicator of the reflectivity shifts on a spiral curve inward toward adaptation point when it approaches the generator.

The reflectivity can be measured by determining the incident and reflected wave parts with the help of a directional coupler (see above) ahead of the point of reflection. With a circuit for sound- and flash-ranging station, in which a probe is directed through horizontally divided circuits, the amplitude path can be measured in the horizontal direction of the circuit.

## 1.6 HOLLOW WAVEGUIDE (H- OR TE-WAVES AND E- OR TM-WAVES)

### 1.6.1 Two-Wire Circuit

If the distance between the conductors of a two-wire circuit discussed in the section on L-waveguides is greater than half the free space wavelength, then in addition to the TEM-mode other field modes are possible.

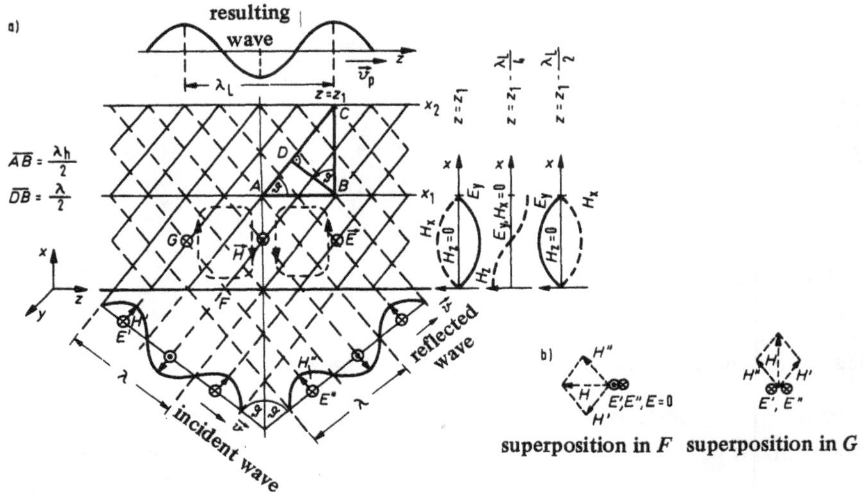


**Fig. 1.16** Reflection of a plane wave in a conducting plane

In order to better understand these field forms, we will first examine the sloping pitch of a plane wave on an ideally conductive ($\kappa \to \infty$), unlimited metal wall (at $\kappa = 0$ in Fig. 1.16a). The medium above the plate is nonconductive and possesses the material constants $\epsilon_r$ and $\mu_r$.

The incident wave ($E', H'$ singled out because of more clarity owing to the underside of the plate) is polarized in the $y$-direction; the incident plane is the $xz$-plane. The incident angle of the wave is equal to $\vartheta$ and its propagation speed is $v$. The propagation is assumed to be $\gamma = j\beta = j2\pi/\lambda$. The wave does not penetrate the metal plate assumed to be ideally conductive: it is completely reflected at that point (components $E'', H''$). The resulting wave field can be obtained by the superposition of the fields of incident and reflected waves. Here the limit conditions $E_t, H_n = 0$ must be satisfied. For the reflected wave, the reflection law, incident angle = angle of reflection, is valid; it propagates at the same speed $v$.

For the field components of incident and reflected waves whose directions of propagation show parts in the $z$- and $y$-direction, the following holds true:

incident wave: (1.126)

$$\underline{H}'_x = \hat{H} \cdot e^{-j\beta(z \cdot \sin\vartheta - x \cdot \cos\vartheta)} \cdot e^{j\omega t}$$

$$\underline{E}'_y = Z_F \cdot \hat{H} \cdot \sin\vartheta \cdot e^{-j\beta(z \cdot \sin\vartheta - x \cdot \cos\vartheta)} \cdot e^{j\omega t} \tag{1.126}$$

$$\underline{E}'_z = Z_F \cdot \hat{H} \cdot \cos\vartheta \cdot e^{-j\beta(z \cdot \sin\vartheta - x \cdot \cos\vartheta)} \cdot e^{j\omega t}$$

reflected wave:

$$\underline{H}''_x = \hat{H} \cdot e^{-j\beta(z \cdot \sin\vartheta + x \cdot \cos\vartheta)} \cdot e^{j\omega t}$$

$$\underline{E}''_y = Z_F \cdot \hat{H} \cdot \sin\vartheta \cdot e^{-j\beta(z \cdot \sin\vartheta + x \cdot \cos\vartheta)} \cdot e^{j\omega t} \tag{1.127}$$

$$\underline{E}''_z = -Z_F \cdot \hat{H} \cdot \cos\vartheta \cdot e^{-j\beta(z \cdot \sin\vartheta + x \cdot \cos\vartheta)} \cdot e^{j\omega t}.$$

Summing the partial waves yields for the resulting wave

$$\underline{H}_x = \underline{H}'_x + \underline{H}''_x = 2\hat{H} \cdot \cos(\beta x \cdot \cos\vartheta) \cdot e^{-j\beta z \cdot \sin\vartheta} \cdot e^{j\omega t}$$

$$\underline{E}_y = \underline{E}'_y + \underline{E}''_y = 2Z_F \cdot \hat{H} \cdot \sin\vartheta \cdot \cos(\beta x \cos\vartheta) \cdot e^{-j\beta z \cdot \sin\vartheta} \cdot e^{j\omega t} \tag{1.128}$$

$$\underline{E}_z = \underline{E}'_z + \underline{E}''_z = 2jZ_F \cdot \hat{H} \cdot \cos\vartheta \cdot \sin(\beta x \cdot \cos\vartheta) \cdot e^{-j\beta z \cdot \sin\vartheta} \cdot e^{j\omega t}.$$

The resulting wave possesses the characteristics of a wave propagating in the $z$-direction and those of a standing wave in the $y$-direction. The phase constant $\beta_h$ and, therefore, the wavelength $\lambda_h = 2\pi/\beta_h$ of the resulting wave propagating in the $z$-direction have changed with regard to the strength of the incident wave: one takes from Eq. (1.128)

$$\beta_h = \frac{2\pi}{\lambda_h} = \beta \cdot \sin\vartheta \tag{1.129}$$

or

$$\lambda_h = \frac{\lambda}{\sin\vartheta} . \tag{1.130}$$

As a function of $y$, the amplitudes of the field components change periodically corresponding to the cosine function or the sine function with the argument $\beta \cdot \kappa \cos\vartheta$. Periodically repeating planes are created in the distances

$$x = \frac{m\pi}{\beta \cdot \cos\vartheta} = \frac{m\lambda}{2 \cdot \cos\vartheta} \qquad m = 1, 2, 3 \ldots \tag{1.131}$$

for oscillation nodes or anti-nodes of the field components as in the case of standing waves. With a vertical descent of the primary wave with $\vartheta = 0$, wave propagation no longer takes place in the $z$-direction ($\beta_h = 0$). The resulting wave is a pure standing wave.

These relationships can be clearly seen by the geometry of the resulting wave fields. To this end, the wave fronts, which are perpendicular to the directions of propagation of the partial waves, are plotted in distances of $\lambda/4$ ($\lambda$ = wavelength of the partial waves) as shown in Fig. 1.16. The vector diagram in Fig. 1.16 for the point $F$ on the metal plate explains the realization of the polarization direction of the resulting wave. The electrical field vector is directed parallel to the wall. In order that the resulting electrical field intensity at the plate equals zero, the $E$-vector of the reflected wave must be directed inversely, the same as in the case of the incident wave: $\vec{E}'' = -\vec{E}'$. We separate the vectors of the magnetic field intensity into both appropriate components $H_x$ and $H_z$. From the condition that the normal components of the resulting magnetic field intensity must disappear at the reflecting metal wall, the reflection follows with $\vec{H}_x'' = -\vec{H}_x'$ and $\vec{H}_z' = -\vec{H}_z'$.

We can find the resulting field intensities at the individual points by adding the parts of incident and reflected waves (maximum + maximum → maximum, minimum + minimum → minimum, maximum + minimum → 0, 0 + 0 → 0). As an example of this vectorial addition, we can refer to Fig. 1.16b: thus, at point $F$ the resulting electrical field intensity becomes zero, the magnetic field intensity becomes maximal and directed opposite to the $z$-direction. At point $G$ the superposition yields a maximum electrical field intensity opposite to the $y$-direction and a maximum magnetic field intensity in the $x$-direction. The lines of force of the resulting magnetic field form (as outlined in Fig. 1.16a) closed curves parallel to the $xz$-plane. The resulting electrical lines of force are perpendicular to the magnetic lines of force.

The resulting field formed by the superposition of the incident and reflected partial plane waves at angle $\vartheta$ shows, in accordance with Eq. (1.131) for planes at distances of

$$x = \frac{m \cdot \lambda}{2 \cdot \cos\vartheta} \qquad m = 1, 2, 3, \ldots$$

with (1.132)

$$\lambda = \frac{\lambda_0}{\sqrt{\mu_r \varepsilon_r}}$$

$\lambda_0 = c/f$ = free space wavelength, from the metal wall (to be read, for example, in the triangle BCD), the same field relationships as on the plate.

If one attaches a second, ideally conductive metal wall in these planes, then the field distribution is maintained between the plates. We can conceive that a sloping, incident plane wave propagates through reflections on the metal walls at the interface.

Figure 1.16a reveals that the superposition of the two waves propagating at angle $\vartheta$ yields a wave propagating in the $z$-direction with a changed wavelength $\lambda_h$. Parallel to the metal wall are locations of maximum oscillation amplitude (oscillation anti-nodes) for the field components and places of decreasing amplitude (oscillation nodes), i.e., characteristics of a standing wave. In the right half of Fig. 1.16a the paths of the field amplitudes depending on $\kappa$ for three $z$-values at intervals separated by $\lambda_h/4$ at a given point in time, or for three consecutive distributions separated by $T/4$ at location $z_1$, are outlined.

The field components of $H_x$ and $H_z$, which are perpendicular to one another, out of phase by 90° with respect to one another, and the magnetic field is, therefore, elliptically polarized.

If the incident angle $\vartheta = 0$, then wave propagation no longer takes place; the superposition of an incident wave with a wave approaching from the opposite direction results in a pure standing wave in the $y$-direction. With the assumed direction of polarization of the incident plane wave, we obtain a resulting field whose electrical field vector is transverse to the direction of propagation and whose magnetic field shows a component in the $z$-direction. We designate such a wave as an *H-wave* or a *transverse electrical (TE-) wave* because of the appearance of an $H$-component in the direction of propagation.

If we proceed to examine a plane wave approaching a metal wall at a slope from a polarization, in which the electrical field vector lies in the $xz$-plane, then we obtain a resulting wave with an electrical field that has one component in the direction of propagation and a pure transverse magnetic field. Such waves are designated as *E-waves* or *transverse magnetic (TM-) waves.*

For the relationship between the wavelength $\lambda_h$ of the resulting wave in the $z$-direction and the wavelength $\lambda$ of the partial waves, the following is valid for the propagation of an $H$-wave as well as for the propagation of an $E$-wave in the two-wire circuit, according to Eq. (1.130),

$$\lambda_h = \frac{2\pi}{\beta_h} = \frac{\lambda}{\sin\vartheta} \tag{1.133}$$

(to be read, for example in triangle ABD).

When we insert $\kappa = a$ for the plate distance in Eq. (1.132), we obtain

$$\lambda_h = \frac{2\pi}{\beta_h} = \frac{\lambda}{\sqrt{1-\cos^2\vartheta}} = \frac{\lambda}{\sqrt{1-\left(\frac{m\lambda}{2a}\right)^2}} \tag{1.134}$$

and for the phase constant

$$\beta_h = \frac{2\pi}{\lambda_h} = \frac{2\pi}{\lambda}\cdot\sqrt{1-\left(\frac{m\lambda}{2a}\right)^2}\,. \tag{1.135}$$

Wave propagation is only possible if $\beta_h$ is real. This is the case for

$$\lambda < \frac{2a}{m} \quad \text{or} \quad \lambda_0 < \frac{2a}{m}\sqrt{\mu_r\varepsilon_r}\,. \tag{1.136}$$

We designate the expression

$$\lambda_c = \frac{c}{f_c} = \frac{2a}{m}\sqrt{\mu_r\varepsilon_r} \tag{1.137}$$

as the *limiting wavelength* or the *critical wavelength.*

Expressed in frequency, we obtain the following

$$f = \frac{c}{\lambda_0} > f_c = \frac{m\cdot c}{2a\cdot\sqrt{\mu_r\varepsilon_r}}\,. \tag{1.138}$$

The frequency $f_c$, above which an $E$- or $H$-wave is possible, is designated as its *limiting* or *critical frequency.*

For $f < f_c$ or $\lambda_0 < \lambda_c$, $\beta_h$ becomes the imaginary part. Therefore, the propagation factor $e^{-j\beta_h \cdot z}$ becomes real; which is to say that long wave propagation is no longer possible. The fields are attenuated in the $z$-direction in proportion to $-j\beta_h$.

Generally, we speak of an $E$- or *H-field mode,* and designate it, if it is capable of propagation ($f > f_c$), as a *wave mode,* and as an *attenuation mode* if it is less than $f_c$.

For $\lambda_0 = \lambda_c$, $\cos\vartheta = 1$, which means $\vartheta = 0°$. According to the model with the reflection of plane waves at the metal plates, this signifies a vertical descent and therefore no propagation for the resulting wave (standing wave). For $\lambda_0 \ll \lambda_c$, $\vartheta \to 0$, the wave propagates in the $z$-direction without reflections; it becomes $\lambda_h \to \lambda_0$.

The propagation of $L$-waves is possible throughout the entire frequency range. Equations (1.134) and (1.135) yield with Eqs. (1.132) and (1.137) an $E$- or $H$-wave for the *wavelength*

$$\lambda_h = \frac{\lambda_0}{\sqrt{\mu_r \varepsilon_r}} \cdot \frac{1}{\sqrt{1 - \left(\frac{\lambda_0}{\lambda_c}\right)^2}} = \frac{\lambda_c}{\sqrt{\mu_r \varepsilon_r}} \cdot \frac{1}{\sqrt{\left(\frac{\lambda_c}{\lambda_0}\right)^2 - 1}} \tag{1.139}$$

and its *phase constant*

$$\beta_h = \frac{2\pi}{\lambda_0} \sqrt{\mu_r \varepsilon_r} \cdot \sqrt{1 - \left(\frac{\lambda_0}{\lambda_c}\right)^2} = \frac{2\pi}{\lambda_c} \sqrt{\mu_r \varepsilon_r} \cdot \sqrt{\left(\frac{\lambda_c}{\lambda_0}\right)^2 - 1} \; . \tag{1.140}$$

For the *attenuation constant in the suppression range* $f < f_c$ as a result of reflection, we obtain

$$\alpha_h = \mathrm{j}\beta_h = \frac{2\pi}{\lambda_c} \sqrt{\mu_r \varepsilon_r} \cdot \sqrt{1 - \left(\frac{\lambda_c}{\lambda_0}\right)^2} \; . \tag{1.141}$$

The functions

$$\frac{\beta_h}{\frac{2\pi}{\lambda_c} \cdot \sqrt{\mu_r \epsilon_r}} \quad \text{and} \quad \frac{\alpha_h}{\frac{2\pi}{\lambda_c} \cdot \sqrt{\mu_r \epsilon_r}}$$

respectively, are graphically represented in Fig. 1.17.

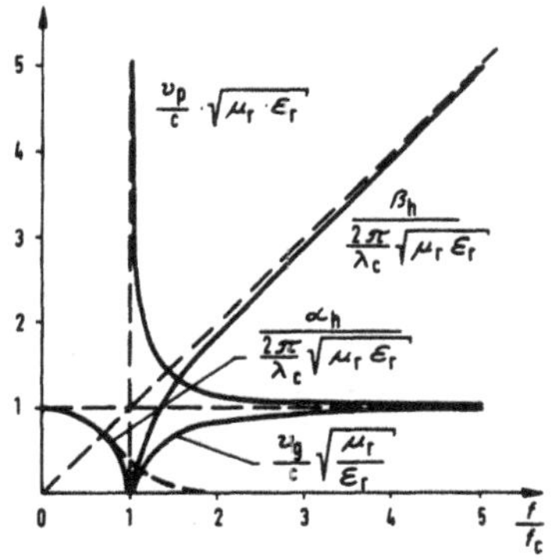


**Fig. 1.17** Normalized representation of phase and reflection attenuation constants and phase and group velocity for a hollow waveguide

For the *phase velocity* at which the field pattern of an $E$- or $H$-wave travels within the two-wire circuit, the following holds true:

$$v_p = \frac{\omega}{\beta_h} = \lambda_h \cdot f = \frac{\lambda}{\sin\vartheta} \cdot f = \frac{c}{\sqrt{\mu_r \varepsilon_r} \cdot \sin\vartheta} = \frac{c}{\sqrt{\mu_r \varepsilon_r}} \frac{1}{\sqrt{1 - \left(\frac{\lambda_0}{\lambda_c}\right)^2}} . \quad (1.142)$$

The *group velocity* at which the transportation of energy or the propagation of a long wave signal in the two-wire circuit advances is obtained as

$$v_g = \frac{d\omega}{d\beta_h} = \frac{c}{\sqrt{\mu_r \varepsilon_r}} \cdot \sin\vartheta = \frac{c}{\sqrt{\mu_r \varepsilon_r}} \cdot \sqrt{1 - \left(\frac{\lambda_0}{\lambda_c}\right)^2} . \quad (1.143)$$

The dependence of the expressions $v_p/c\sqrt{(\mu_r \epsilon_r)}$ and $v_g/c\sqrt{(\mu_r \epsilon_r)}$ upon $f/f_c$ is represented in Fig. 1.17. The *characteristic wave impedances,* which are designated as the relationship between the electrical and magnetic cross field intensities for waves propagating at a point in the field, are different for $H$-wave and $E$-wave propagation.

For $H$-waves the following is valid

$$Z_{F,H} = \frac{Z_E}{\sin\vartheta} = \frac{Z_F}{\sqrt{1 - \left(\frac{\lambda_0}{\lambda_c}\right)^2}} \quad (1.144)$$

and for $E$-waves

$$Z_{F,E} = Z_F \cdot \sin\vartheta = Z_F \cdot \sqrt{1 - \left(\frac{\lambda_0}{\lambda_c}\right)^2} \quad (1.145)$$

with

$$Z_F = 120\,\pi \sqrt{\frac{\mu_r}{\varepsilon_r}}\,\Omega .$$

These relations will be derived in sec. 1.6.2 on hollow waveguides.

In the suppression range of the waveguide, the oscillation resistance becomes the imaginary part.

We designate the number $m = 1, 2, 3 \ldots$, which occurs in the equations for the indication of the special field mode: the number $m$ specifies the quantity of sine semicycles or cosine semicycles in the field distribution in the $x$-direction along the segment $a$ between the plates (compare with Fig. 1.16a to the right). In the $y$-direction the field components at the assumed ideal conductor do not depend upon location; we express this with a second index 0. Therefore, in the case examined in Fig. 1.16, we obtain by arranging the second plate in the plane $x_1$ = const, an $H_{10}$-field mode, and if we insert this at point $x_2$, then we obtain an $H_{20}$-mode with a repeating and out of phase field distribution.

The characteristics of $E$- and $H$-waves will be more closely examined in connection with hollow waveguides in the next section.

### 1.6.2 Hollow Waveguides

Propagation of electromagnetic waves is also possible in hollow tubes called hollow waveguides or simply waveguides. If no other indication is made in the following, then metallic hollow waveguides are being referred to. Dielectric hollow waveguides will be dealt with in the sec. 1.8 on surface waves.

The possibility of the existence of $H$-waves in the circular hollow waveguide was first theoretically proven by J.J. Thomson in 1893. J.W. Stutt (Lord Rayleigh) dealt subsequently (1897) with $H$- and $E$-waves in circular and rectangular hollow waveguides. A.Becker was the first to experimentally prove the existence of hollow waveguide waves in 1902.

In the following two sections we will deal with hollow waveguides with rectangular and curved cross sections which are of particular technical importance.

#### *1.6.2.1 Rectangular Hollow Waveguides*

If we shut off the two-wire circuit at the location of the field through two additional conducting walls parallel to the $xz$-plane at a distance $b$, we obtain a rectangular hollow waveguide (Fig. 1.18).

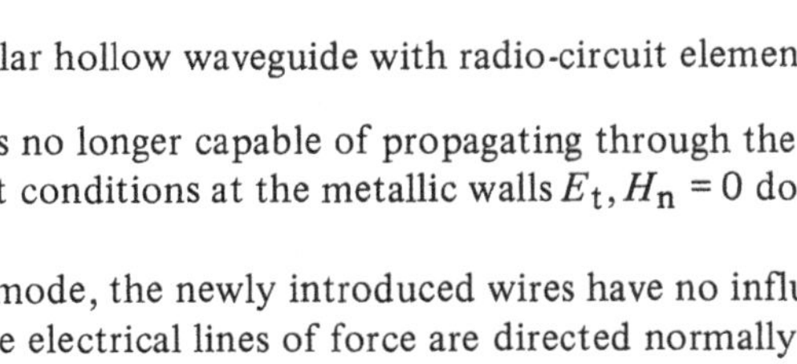

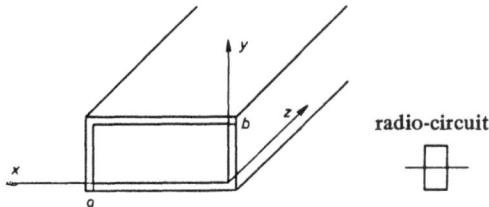

**Fig. 1.18** Rectangular hollow waveguide with radio-circuit element

The TEM-wave is no longer capable of propagating through the hollow waveguide since the limit conditions at the metallic walls $E_t, H_n = 0$ do not allow this.

In an $H_{m0}$-field mode, the newly introduced wires have no influence because, at their location, the electrical lines of force are directed normally and the magnetic lines of force are directed tangentially.

If, by examining the inclined reflection at a metal wall, we choose the incident wave polarization such that the normal wave lines of the partial plane waves also receive in the $y$-direction (incident plane is no longer the $xz$-plane), then the existence of $H_{mn}$- and $E_{mn}$-waves in the rectangular hollow waveguide can also be generally explained by this model (introduced by L. Brillouin).

In order to investigate systematically the field modes possible in the rectangular hollow waveguide, we begin with Maxwell's equations (1.7) and (1.8), for which solutions must be found that satisfy the limit conditions at the assumed ideally conductive inner walls of the hollow waveguide.

The circuit is assumed to be lossless, of great length, and homogeneous, which means with the same size and material characteristics in every plane of cross section. $\epsilon = \epsilon_r \cdot \epsilon_0$ and $\mu = \mu_r \cdot \mu_0$ are the dielectric and permeability constants of the medium which uniformly fills the waveguide's interior. Most often $\epsilon_r = \mu_r = 1$. The part of the conductive current in Eq. (1.7) cancels out ($\kappa = 0$) in the circuit's interior.

The field intensities can no longer remain constant within the cross-section area of the circuit, so that for the general case of $\partial/\partial x$, we must assume $\partial/\partial y \neq 0$. The dependence of the field components upon propagation in the $z$-direction is established by complex notation with $e^{-j\beta_h \cdot z}$, so that the partial derivatives according to $z$ yield results through multiplication by $-j\beta_h$. The time dependence is assumed to be harmonic with the complex time factor $e^{j\omega t}$.

The solution of Maxwell's equation for the individual field components yields

$$\frac{\partial \underline{E}_z}{\partial y} + j\beta_h \cdot \underline{E}_y = -j\omega\mu \underline{H}_x \tag{1.146}$$

$$-j\beta_h \underline{E}_x - \frac{\partial \underline{E}_z}{\partial x} = -j\omega\mu \underline{H}_y \tag{1.147}$$

$$\frac{\partial \underline{E}_y}{\partial x} - \frac{\partial \underline{E}_x}{\partial y} = -j\omega\mu \underline{H}_z \tag{1.148}$$

$$\frac{\partial \underline{H}_z}{\partial y} + j\beta_h \cdot \underline{H}_y = j\omega\varepsilon \underline{E}_x \tag{1.149}$$

$$-j\beta_h \underline{H}_x - \frac{\partial \underline{H}_z}{\partial x} = j\omega\varepsilon \underline{E}_y \tag{1.150}$$

$$\frac{\partial \underline{H}_y}{\partial x} - \frac{\partial \underline{H}_x}{\partial y} = j\omega\varepsilon \underline{E}_z \tag{1.151}$$

It is possible to represent the cross field components through the horizontal field components $\underline{E}_z$, $\underline{H}_z$, as was shown for the component $\underline{E}_x$:

The insertion of Eq. (1.147) into Eq. (1.149) yields

$$j\omega\varepsilon \underline{E}_x = \frac{\partial \underline{H}_z}{\partial y} - \frac{\beta_h}{\omega\mu}\left(-j\beta_h \underline{E}_x - \frac{\partial \underline{E}_z}{\partial x}\right). \tag{1.152}$$

The solution according to $\underline{E}_x$ yields

$$\underline{E}_x = -\frac{j}{\omega^2\mu\varepsilon - \beta_h^2}\left(\beta_h \frac{\partial \underline{E}_z}{\partial x} + \omega\mu \frac{\partial \underline{H}_z}{\partial y}\right). \tag{1.153}$$

For the remaining field components, we obtain in the same manner

$$\underline{E}_y = \frac{j}{\omega^2\mu\varepsilon - \beta_h^2}\left(-\beta_h \frac{\partial \underline{E}_z}{\partial y} + \omega\mu \frac{\partial \underline{H}_z}{\partial x}\right) \tag{1.154}$$

$$\underline{H}_x = \frac{j}{\omega^2\mu\varepsilon - \beta_h^2}\left(\omega\varepsilon \frac{\partial \underline{E}_z}{\partial y} - \beta_h \frac{\partial \underline{H}_z}{\partial x}\right) \tag{1.155}$$

$$\underline{H}_y = -\frac{j}{\omega^2\mu\varepsilon - \beta_h^2}\left(\omega\varepsilon \frac{\partial \underline{E}_z}{\partial x} + \beta_h \frac{\partial \underline{H}_z}{\partial y}\right). \tag{1.156}$$

The components $\underline{E}_z$ and $\underline{H}_z$ are independent of one another.

If we set $H_z = 0$, then we obtain $E$-waves (TM-waves): the differential equation (wave equation) for a field component is obtained by eliminating the remaining components from Eqs. (1.146) to (1.151). This is the same for all field components,* and is written for $\underline{E}_z$ in the form

$$\frac{\partial^2 \underline{E}_z}{\partial x^2} + \frac{\partial^2 \underline{E}_z}{\partial y^2} + (\omega^2\mu\varepsilon - \beta_h^2)\,\underline{E}_z = 0\,. \tag{1.157}$$

This partial differential equation can be solved by separating the variable quantities with the help of a product statement $\underline{E}_z = E_0 \cdot f(x) \cdot g(y) \cdot e^{j(\omega t - \beta_h \cdot z)}$. Concering the limit condition that the tangential component $\underline{E}_z$ at the metal surfaces must disappear for $x = 0, x = a$ and $y = 0, y = b$, we find the solution

$$\underline{E}_z = E_0 \cdot \sin\frac{m\pi x}{a} \cdot \sin\frac{n\pi y}{b} \cdot e^{j(\omega t - \beta_h \cdot z)}\,. \tag{1.158}$$

If we insert the statement for $\underline{E}_z$ into the differential equation, then we obtain the relation for the phase constant $\beta_h$ or the wavelength $\lambda_h$ of the hollow waveguide, as follows:

$$\beta_h^2 = \left(\frac{2\pi}{\lambda_h}\right)^2 = \omega^2\mu\varepsilon - \left(\frac{m\pi}{a}\right)^2 - \left(\frac{n\pi}{b}\right)^2. \tag{1.159}$$

---

[1] Vectorally Analytical Solution Path:
rot $\underline{\vec{E}} = -j\omega\mu\underline{\vec{H}}$, rot $\underline{\vec{H}} = j\omega\varepsilon\underline{\vec{E}}$, div $\underline{\vec{E}} = 0$, div $\underline{\vec{H}} = 0$. With rot rot $\underline{\vec{A}}$ = grad div $\underline{\vec{A}} - \nabla^2\underline{\vec{A}}$:
rot rot $\underline{\vec{E}} = \omega^2\mu\varepsilon\underline{\vec{E}}$, rot rot $\underline{\vec{H}} = \omega^2\mu\varepsilon\underline{\vec{H}}$ or $\nabla^2\underline{\vec{E}} + \omega^2\mu\varepsilon\underline{\vec{E}} = 0$, $\nabla^2\underline{H} + \omega^2\mu\varepsilon\underline{\vec{H}} = 0$;
$$\nabla^2 = \frac{\partial^2}{\partial x^2} + \frac{\partial^2}{\partial y^2} + \frac{\partial^2}{\partial z^2}\,.$$

If we introduce the phase constant $\beta$ for plane waves, according to Eqs. (1.37) and (1.43),

$$\beta = \frac{2\pi}{\lambda} = \frac{2\pi}{\dfrac{\lambda_0}{\sqrt{\mu_r \varepsilon_r}}} = \omega\sqrt{\mu\varepsilon}\ , \tag{1.160}$$

$\lambda_0 = c/f$ = free-space wavelength, then Eq. (1.159) can be written as

$$\beta_h^2 = \beta^2 - \beta_c^2 \tag{1.161}$$

with

$$\beta_c = \frac{2\pi}{\dfrac{\lambda_c}{\sqrt{\mu_r \varepsilon_r}}} = \sqrt{\left(\frac{m\pi}{a}\right)^2 + \left(\frac{n\pi}{b}\right)^2}\ . \tag{1.162}$$

Wave propagation within the hollow waveguide is only possible for a real phase $\beta_h$, which is to say, for

$$\beta_h^2 > 0 \qquad \text{or} \qquad \beta > \beta_c\ . \tag{1.163}$$

With $\beta = 2\pi/\lambda_0\sqrt{(\mu_r \epsilon_r)}$, the condition for wave propagation in the hollow waveguide can also be written as

$$\lambda_0 < \frac{2\pi\sqrt{\mu_r \varepsilon_r}}{\beta_c} = \frac{2\sqrt{\mu_r \varepsilon_r}}{\sqrt{\left(\dfrac{m}{a}\right)^2 + \left(\dfrac{n}{b}\right)^2}}\ . \tag{1.164}$$

We designate the expression

$$\lambda_c = \frac{2\sqrt{\mu_r \varepsilon_r}}{\sqrt{\left(\dfrac{m}{a}\right)^2 + \left(\dfrac{n}{b}\right)^2}} \tag{1.165}$$

as the *limiting wavelength* or the *critical wavelength.* This is

$$\lambda_c = \frac{c}{f_c}\ . \tag{1.166}$$

This means that the critical wavelength is given as the wavelength possessed by a plane wave in free space and at the frequency $f_c$.

With $f = c/\lambda_0$ the condition for wave propagation can be expressed as

$$f > f_c\ . \tag{1.167}$$

Therefore, wave propagation is possible in the hollow waveguide only for frequencies above the *limiting frequency* or *critical frequency,* or rather for wavelengths smaller than the critical wavelength $\lambda_c$. The hollow waveguide has high-pass transmission characteristics. The magnitudes $f_c$ and $\lambda_c$, corresponding to their $m$- and $n$-values, are different for the various field modes. With Eqs. (1.161) and (1.160), $\beta_h$ and $\lambda_n$ can be expressed as

$$\beta_h = \sqrt{\beta^2 - \beta_c^2} = \beta \cdot \sqrt{1 - \left(\frac{\beta_c}{\beta}\right)^2} = \frac{2\pi\sqrt{\mu_r \varepsilon_r}}{\lambda_0} \cdot \sqrt{1 - \left(\frac{\lambda_0}{\lambda_c}\right)^2} = \tag{1.168}$$

$$= \frac{2\pi\sqrt{\mu_r \varepsilon_r}}{\lambda_c} \cdot \sqrt{\left(\frac{\lambda_c}{\lambda_0}\right)^2 - 1}$$

and

$$\lambda_h = \frac{2\pi}{\beta_h} = \frac{\lambda_0}{\sqrt{\mu_r \varepsilon_r}} \cdot \frac{1}{\sqrt{1 - \left(\frac{\lambda_0}{\lambda_c}\right)^2}} = \frac{\lambda_c}{\sqrt{\mu_r \varepsilon_r}} \cdot \frac{1}{\sqrt{\left(\frac{\lambda_c}{\lambda_0}\right)^2 - 1}} . \tag{1.169}$$

We obtain, therefore, the same relations as in the case of the two-wire circuit, where, for the critical wavelength, Eq. (1.165) must now be chosen instead of Eq. (1.137). Equation (1.137) arises from Eq. (1.165) with $n = 0$.

For $f < f_c$, $\beta_h^2 < 0$, therefore, $\beta_h$ becomes the imaginary part. Therefore, the propagation factor $e^{-j\beta_h \cdot z}$ becomes real. The electrical and magnetic fields no longer propagate in the wave form: they oscillate in-phase and decrease exponentially corresponding to $e^{-\alpha_h \cdot z}$ in the $z$-direction with the attenuation constant

$$\alpha_h = j\beta_h = \sqrt{\beta_c^2 - \beta^2} = \frac{2\pi\sqrt{\mu_r \varepsilon_r}}{\lambda_c} \cdot \sqrt{1 - \left(\frac{\lambda_c}{\lambda_0}\right)^2} . \tag{1.170}$$

Here we are dealing with a field attenuation due to the reflection, in contrast to attenuation through wall current losses which will be dealt with later. This aperiodic behavior of hollow waveguides below their critical frequency can be used in the construction of attenuator pads and filters.

For the path of the frequency-dependent phase and reflection attenuation constants, the same curves are valid according to Fig. 1.17, as in the case of the two-wire circuit.

The different field forms possible according to the values $m$ and $n$ are generally designated as *field modes.* If these are capable of propagation ($f > f_c$), they are called *wave modes* (periodic change of field in the direction of propagation), and in the aperiodic case ($f < f_c$), *attenuation modes* (exponential decrease of the field magnitudes in the $z$-direction).

The field components of the $\underline{E}$-field modes which are still missing can be obtained from Eqs. (1.153) to (1.156). Thereby, we obtain for the components of the *E-field modes in the rectangular hollow waveguide*

$$\underline{E}_z = E_0 \cdot \sin\frac{m\pi x}{a} \cdot \sin\frac{n\pi y}{b} \cdot e^{j(\omega t - \beta_h \cdot z)} \tag{1.171}$$

$$\underline{E}_x = E_0 \cdot \frac{\beta_h}{\beta_c^2} \cdot \frac{m\pi}{a} \cdot \cos\frac{m\pi x}{a} \cdot \sin\frac{n\pi y}{b} \cdot e^{j(\omega t - \beta_h \cdot z - \pi/2)} \tag{1.172}$$

$$\underline{E}_y = E_0 \cdot \frac{\beta_h}{\beta_c^2} \cdot \frac{n\pi}{b} \cdot \sin\frac{m\pi x}{a} \cdot \cos\frac{n\pi y}{b} \cdot e^{j(\omega t - \beta_h \cdot z - \pi/2)} \tag{1.173}$$

$$\underline{H}_z = 0 \tag{1.174}$$

$$\underline{H}_x = -\frac{\underline{E}_y}{Z_{F,E}} \tag{1.175}$$

$$\underline{H}_y = \frac{\underline{E}_x}{Z_{F,E}}\,. \tag{1.176}$$

In the lossless hollow waveguide with purely propagating waves, the relationship between the transverse components of the electrical field intensity vector and the transverse component of the magnetic field intensity (which stands perpendicular to the former) is real and of equal magnitude throughout. This relationship is designated as the *characteristic wave impedance.* For $E$-waves the following holds true:

$$\begin{aligned} Z_{F,E} &= Z_0 \cdot \sqrt{\frac{\mu_r}{\varepsilon_r}} \cdot \frac{\beta_h \cdot \lambda_0}{2\pi\sqrt{\mu_r\varepsilon_r}} = \frac{\beta_h}{\omega \cdot \varepsilon_r\varepsilon_0} = \\ &= Z_0 \cdot \sqrt{\frac{\mu_r}{\varepsilon_r}} \cdot \sqrt{1 - \left(\frac{\lambda_0}{\lambda_c}\right)^2} \end{aligned} \tag{1.177}$$

with $Z_0 = \sqrt{(\mu_0/\epsilon_0)} = 120\pi\Omega$ = characteristic wave impedance of a plane wave in free space.

If we proceed from the assumption $E_z = 0$ in solving the set of equations (1.146) to (1.151) or (1.153) to (1.156), then we obtain *H-(TE-) waves:*

For the field components, a differential equation in the form of Eq. (1.157) is once again valid. The field components are now derived from the horizontal field intensity $\underline{H}_z$. Since $\underline{H}_z$ is a tangential component with respect to the waveguide wall and does not disappear, the limit conditions for the electrical field intensity apply when establishing the equation for $\underline{H}_z$. $E_y$ must equal 0 for $x = 0$ and $x = a$, and $E_x$ must equal 0 for $y = 0$ and $y = b$. Therefore, we obtain as the limit conditions to be met from Eqs. (1.154) and (1.153) with $\underline{E}_z = 0$: $\partial\underline{H}_z/\partial x = 0$ for $x = 0$ and $x = a$, and $\partial\underline{H}_z/\partial y = 0$ for $y = 0$ and $y = b$. The remaining field components can be derived from $\underline{H}_z$ with the help of Eqs. (1.153) to (1.156). Therefore, the components of the *H-field modes in the rectangular hollow waveguide* are obtained as

$$\underline{H}_z = H_0 \cdot \cos\frac{m\pi x}{a} \cdot \cos\frac{n\pi y}{b} \cdot e^{j(\omega t - \beta_h \cdot z)} \tag{1.178}$$

$$\underline{H}_x = H_0 \cdot \frac{\beta_h}{\beta_c^2} \cdot \frac{m\pi}{a} \cdot \sin\frac{m\pi x}{a} \cdot \cos\frac{n\pi y}{b} \cdot e^{j(\omega t - \beta_h \cdot z + \pi/2)} \tag{1.179}$$

$$\underline{H}_y = H_0 \cdot \frac{\beta_h}{\beta_c^2} \cdot \frac{\pi}{b} \cdot \cos\frac{m\pi x}{a} \cdot \sin\frac{n\pi y}{b} \cdot e^{j(\omega t - \beta_h \cdot z + \pi/2)} \tag{1.180}$$

$$\underline{E}_z = 0 \tag{1.181}$$

$$\underline{E}_x = \underline{H}_y \cdot Z_{F,H} \tag{1.182}$$

$$\underline{E}_y = -\underline{H}_x \cdot Z_{F,H} \tag{1.183}$$

with the characteristic wave impedance

$$Z_{F,H} = \frac{Z_0 \cdot \sqrt{\frac{\mu_r}{\varepsilon_r}} \cdot 2\pi \cdot \sqrt{\mu_r \varepsilon_r}}{\beta_h \cdot \lambda_0} = \frac{\omega \cdot \mu_r \mu_0}{\beta_h} = \frac{Z_0 \cdot \sqrt{\frac{\mu_r}{\varepsilon_r}}}{\sqrt{1 - \left(\frac{\lambda_0}{\lambda_c}\right)^2}} \tag{1.184}$$

with $Z_0 = \sqrt{\frac{\mu_0}{\varepsilon_0}} = 120\,\pi\Omega$ .

For the path of $Z_{F,E}/Z_0 \cdot \sqrt{(\mu_r/\epsilon_r)}$ and $Z_{F,H}/Z_0 \cdot \sqrt{(\mu_r/\epsilon_r)}$ the representation of $v_g/c\,\sqrt{(\mu_r\epsilon_r)}$ or $v_p/c\,\sqrt{(\mu_r\epsilon_r)}$ can be used according to Fig. 1.17.

For the phase and attenuation constants in the suppression range, the waveguide wavelengths, and the *H*-wave critical wavelengths, the same relations (1.168), (1.170), (1.169), and (1.165) are valid for *E*-waves.

For the *phase velocity* $v_p$ of an *E*- or *H*-wave propagating in a hollow waveguide, whereby the field pattern shifts in the *z*-direction, we obtain with Eq. (1.168)

$$v_p = \frac{\omega}{\beta_h} = \frac{\omega}{2\pi \cdot \sqrt{\mu_r \varepsilon_r}} \cdot \frac{1}{\sqrt{1 - \left(\frac{\lambda_0}{\lambda_c}\right)^2}} = \frac{c}{\sqrt{\mu_r \varepsilon_r}} \cdot \frac{1}{\sqrt{1 - \left(\frac{\lambda_0}{\lambda_c}\right)^2}} . \tag{1.185}$$

For the *group velocity* $v_g$ where a small group of assorted frequency components propagates in the hollow waveguide (i.e., an information rate per time or energy flow) we obtain after differentiation of Eq. (1.168) and with $\lambda_0/\lambda_c = \omega_c/\omega$

$$v_g = \frac{d\omega}{d\beta_h} = \frac{c}{\sqrt{\mu_r \varepsilon_r}} \cdot \sqrt{1 - \left(\frac{\omega_c}{\omega}\right)^2} = \frac{c}{\sqrt{\mu_r \varepsilon_r}} \cdot \sqrt{1 - \left(\frac{\lambda_0}{\lambda_c}\right)^2} . \tag{1.186}$$

The product of the phase and group velocities is equal to the speed of light squared.

The dependence of the propagation velocity on frequency, which is designated as the *dispersion,* leads to linear distortions in the case of wideband signal transmission.

The phase and group velocities for the lossless, homogeneous hollow waveguide, together with their phase and reflection attenuation constants, are recorded as a function of frequency for the normalized representation in Fig. 1.17.

At frequencies well above the critical frequency, the propagation relationships approach those of plane waves.

We characterize the various *E*- and *H*-field modes possible in the rectangular hollow waveguide as $E_{mn}$- or $H_{mn}$-modes, whereby the subscripts *m* and *n* are the whole numbers *m* and *n* in the equations of the field components and the critical wavelengths present:

The *first subscript m* characterizes the number of half-cycles of the magnetic (electrical) cross field along the larger inner side (*x*-direction) of the hollow waveguide in the case of $E_{mn}$- ($H_{mn}$-) modes, and the *second subscript n* characterizes the corresponding number of half-cycles along the smaller inner side (*y*-direction).

With increasing frequency, a much greater number of such natural waves can be propagated in the hollow waveguide.

In Fig. 1.19, the paths of electrical and magnetic lines of force in the cross sections and horizontal sections of the circuit are outlined for some field modes of the rectangular hollow waveguide.

The current density $\underline{S}_v = \epsilon \cdot dE/dt = j\omega\underline{E}$ has a forward phase shift of $\omega t = \pi/2$ in relation to the electrical field strength for propagating waves. The shift current lines can thus be obtained by a shift of the electrical lines of force of $\lambda_h/4$ in the direction of propagation; these lines are enclosed by the magnetic lines of force (circulation law).

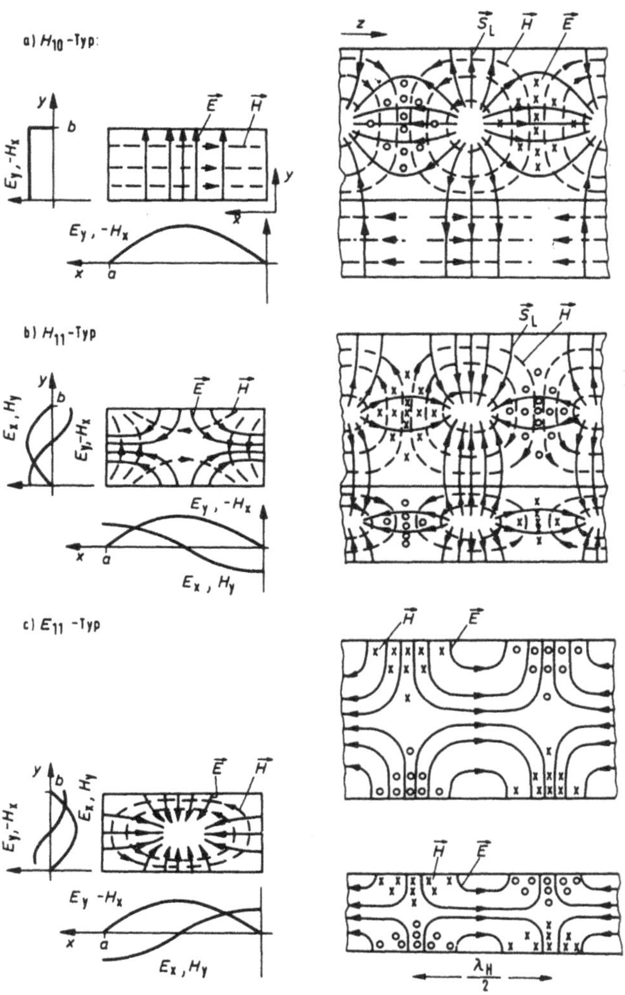


**Fig. 1.19** Field distribution in rectangular hollow waveguides: in (a) and (b) with wall current distribution

The *wall currents* restore the shift currents to closed electrical circuits. The wall currents are proportional to the tangential magnetic field at the wall and perpendicular to it (see skin effect). They flow through a thin layer on the inner side of the waveguide wall. $E$-waves possess only axial wall currents since the magnetic field shows only transverse components.

The energy (lines of force) pattern of $E_{mn}$- and $H_{mn}$-modes for higher power in the cross section can be obtained by putting together the patterns of $E_{11}$- or $H_{11}$-modes $m$-times along the width and $n$-times along the narrow side, whereby the field components in the neighboring ranges are out of phase. The field patterns of an $H_{0i}$-mode correspond to the field pattern of the $H_{i0}$-mode rotated 90°.

In the case of the hollow waveguide fields, the transverse components of the electrical and magnetic fields are also perpendicular to one another.

In the case of $E$-waves, $E_x, E_y$, and $H_y$ oscillate in phase and lag with respect to $E_z$ temporally by 90°, or rather are shifted by $\lambda_h/4$ opposite the $z$-direction relative to $E_z$ • $H_x$ leads $E_z$ by 90° or is shifted by $\lambda_h/4$ in the $z$-direction.

In the case of $H$-waves, $E_x, H_x$, and $H_y$ are in phase and lead relative to $H_z$ by 90°, or rather are shifted by $\lambda_h/4$ in the $z$-direction with respect to $H_z$. $E_y$ lags behind $H_z$ by 90°, or is shifted by $\lambda_h/4$ opposite the $z$-direction relative to $H_z$ [compare with Eqs. (1.171) to (1.176) and (1.178) to (1.183)].

The fact that with wave modes (periodic fields in the $z$-direction) the transverse components of the electrical and magnetic fields are in or out of phase and shifted in phase with respect to the horizontal field components by ±90°, means that the $z$-component of the Poynting vector of the equalizing current $\vec{P}' = \vec{E} \times \vec{H}$ is positively real, whereas its transverse components are imaginary.

In the case of attenuation modes (aperiodic fields in the $z$-direction), the electrical and magnetic transverse field intensities are shifted in phase by $\pi/2$ with respect to one another; with $P_z'$ imaginary.

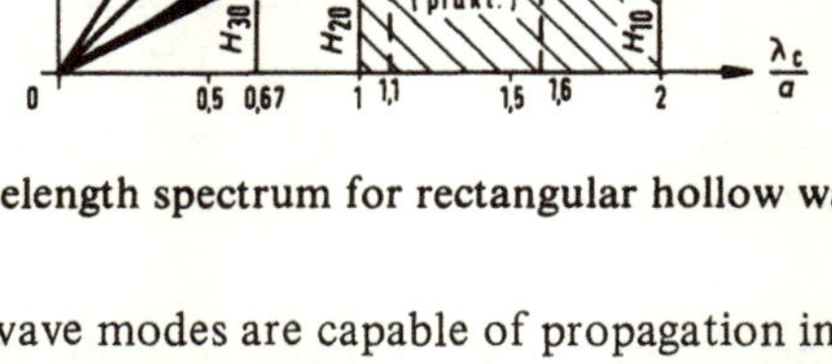

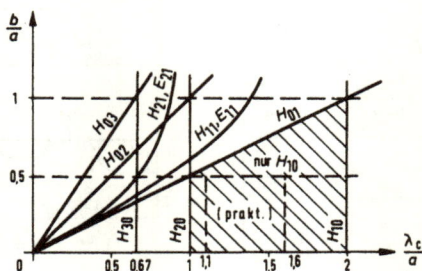

**Fig. 1.20** Critical wavelength spectrum for rectangular hollow waveguides

A survey of which wave modes are capable of propagation in a rectangular hollow waveguide at given dimensions $a, b$ and a given frequency $f = c/\lambda_0$, yields the so-called *critical wavelength spectrum* in Fig. 1.20. In this figure the relationship $b/a$ is plotted as a function of $\lambda_c/a$ for some $m$- and $n$-values as a critical curve according to Eq. (1.165) for $\epsilon_r = \mu_r = 1$. Wave propagation is possible for $\lambda_0/a < \lambda_c/a$.

The $H_{10}$-mode is the field mode with the shortest critical wavelength $\lambda_c = 2a$. It is, therefore, designated as the *primary mode* or the *basic mode.* Only the $H_{10}$-wave is capable of propagating in the shaded region of the diagram given in Fig. 1.20. The surface wave or primary line is of the highest technical significance because of the large frequency range in which it can be stimulated without other kinds of disturbance.

How do we choose the cross-section dimensions of a rectangular hollow waveguide for the transmission of an $H_{10}$-wave in practice?

We are mostly interested in a wide operating frequency range for a given hollow waveguide. This is restricted downwards by the critical frequency of the $H_{10}$-wave ($f_c = c/2a$), and upwards by the critical frequencies of higher field modes ($H_{20}$-mode with $f_c = c/a$ and $H_{01}$-mode with $f_c = c/2b$). Because of the increased attenuation and the radical change in propagation velocity and oscillation resistance of the surface wave when approaching its critical frequency, we will adhere to a corresponding distance.

Disturbances in the geometry of the hollow waveguide, such as contortions, warpings, and transitions, give rise to the formation of additional field modes. To avoid these disturbances, we will adhere to a safety zone with the operating frequency above its critical frequencies as well.

An increase in the length $b \leqslant a$ of the waveguide's narrow side increases the transferable power provided by the disruptive force and lessens the attenuation. However, in this case, the danger of stimulating the $H_{01}$-mode increases.

Therefore, we usually choose

$$\frac{b}{a} = 0{,}5 \qquad \text{and} \qquad 1{,}25\, f_c < f < 1{,}9\, f_c\,. \tag{1.187}$$

The German standard for the rectangular hollow waveguide follows from DIN 47 302, part 1. The specifications for hollow waveguides in the frequency range $f = 320$ MHz to 330 GHz ($H_{10}$-wave) are listed therein.

In addition to the normal profile with $b/a = 0.5$, occasionally flat profiles are used, most often with an aspect ratio $b/a = 1{:}8$ or $1{:}4$. With such narrow-profile waveguides, the transfer power is less, the attenuation higher and the oscillation resistance smaller. We use waveguides with reduced height because of the space saving within instruments, and because of the easier adaptation with lower oscillation resistance to coaxial transmission lines or microstrip lines (for wide-band transmissions), or low resistance circuits (e.g., diodes).

By partially or completely filling the hollow waveguide with dielectric material, the requisite size can be reduced (in the case of filling completely by the factor $1/\sqrt{\epsilon_r}$ at the same frequency $f_c$). With this the attenuation climbs and the transferable power increases.

We can attain an increased frequency range with the hollow ridge waveguide (see above) up to values of 4:1 for the $H_{10}$-mode.

If a hollow waveguide is operated in a frequency range where only the $H_{10}$-wave is capable of propagating, then at inhomogeneous locations along the line only aperiodic fields of higher power occur, which die down quickly at increasing distance from the center of the disturbance and do not deprive the wave of its active power.

$E_{m0}$- and $E_{0n}$-field modes cannot exist in the rectangular hollow waveguide since the $E_z$-component present in $E$-modes is short circuited on either the wide or narrow pair of wires.

In order to stimulate a certain field mode in a hollow waveguide, we form the point of excitation in such a way that it shows field components, as in the case of the waveguide field to be stimulated. The excitation can take place electrically with a terminal parallel to the electrical lines of force of the field mode to be stimulated, or magnetically with a belt loop whose magnetic field is directed in the same way as the field to be stimulated. Therefore, for example, with the internal conductor of a coaxial transmission line projected into the hollow waveguide, we can stimulate an $H_{10}$-field by orientation in the $y$-direction, and an $E_{11}$-field by orientation in the $z$-direction. For better adaptation, matching elements can be installed (see also hollow waveguide transitions).

### *1.6.2.2 Circular Hollow Waveguides*

In order to determine the characteristics of the fields possible within a circular hollow waveguide, we must find the solutions to Maxwell's equations in cylindrical coordinates $\rho, \varphi, z$ (compare with Fig. 1.21) which satisfy the limit conditions.

Similar to that of the field components in the rectangular hollow waveguide, we can obtain a representation of the cross field intensities $\underline{E}_\rho, \underline{E}_\varphi, \underline{H}_\rho$, $\underline{H}_\varphi$ as a function of the horizontal field intensities $\underline{E}_z$ and $\underline{H}_z$ are independent of one another. This leads again to a separation of $E$-waves ($H_z = 0$) and $H$-waves ($E_z = 0$).

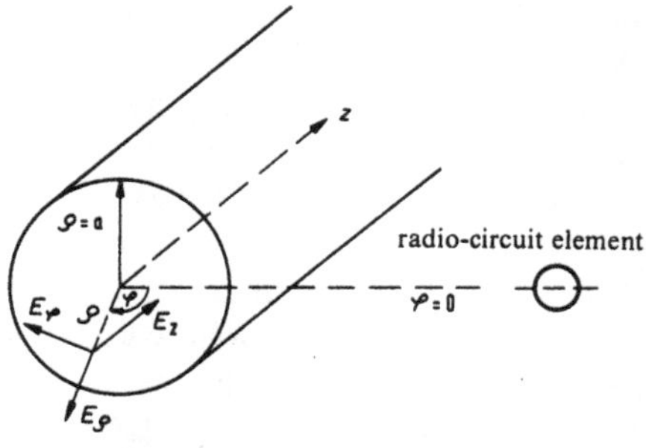


**Fig. 1.21** Circular hollow waveguide with radio-circuit element

The wave equation valid for all field components has the following form in cylindrical coordinates:

$$\frac{\partial^2 \underline{\vec{E}}}{\partial \varrho^2} + \frac{1}{\varrho} \cdot \frac{\partial \underline{\vec{E}}}{\partial \varrho} + \frac{1}{\varrho^2} \cdot \frac{\partial^2 \underline{\vec{E}}}{\partial \varphi^2} + \frac{\partial^2 \underline{\vec{E}}}{\partial z^2} + \omega^2 \mu \varepsilon \underline{\vec{E}} = 0 . \qquad (1.188)$$

First, we shall investigate the $E$-waves with $H_z = 0$.
Proceeding from the component $E_z$, we find for its field components:

$$\underline{E}_z = E_0 \cdot \cos(m\varphi) \cdot J_m(\beta_c \varrho) \cdot e^{j(\omega t - \beta_h \cdot z)} \qquad (1.189)$$

$$\underline{E}_\varrho = E_0 \cdot \frac{\beta_h}{\beta_c} \cos(m\varrho) \cdot J'_m(\beta_c \varrho) \cdot e^{j(\omega t - \beta_h z + \pi/2)} \qquad (1.190)$$

$$\underline{E}_\varphi = E_0 \cdot \frac{\beta_h}{\beta_c^2} \cdot \frac{m}{\varrho} \cdot \sin(m\varphi) \cdot J_m(\beta_c \varrho) \cdot e^{j(\omega t - \beta_h \cdot z - \pi/2)} \qquad (1.191)$$

$$\underline{H}_z = 0 \qquad (1.192)$$

$$\underline{H}_\varrho = - \frac{\underline{E}_\varphi}{Z_{F,E}} \qquad (1.193)$$

$$\underline{H}_\varphi = \frac{\underline{E}_\varrho}{Z_{F,E}} . \qquad (1.194)$$

Thus, $J_m(\beta_c \rho)$ is the Bessel function to the $m$th power from the argument $\beta_c \cdot \rho$. In Fig. 1.22, the functions $J_m(x)$ for $m = 0, 1, 2$, and their derivative functions $J'_m(x)$ are represented graphically.

For the phase constant $\beta_h = 2\pi/\lambda_h$ and the characteristic wave impedance $Z_{F,E}$, the same relations (1.168) and (1.177) are valid as for the rectangular hollow waveguide. The critical wavelength which is now to be inserted in these formulas can be obtained from the following consideration: at the conducting wall of the hollow waveguide, the tangential electrical field intensity components $E_\varphi, E_z$ and the normal magnetic field intensity $H_\rho$ must disappear. As the above equations state, to this end the functions $J_m$ must show zero positions for $\rho = a$. The zero positions are designated by $j_{mn}$, whereby subscript $m$ states the ordinal number of the Bessel function in question and $n$ states the number of the zero position. An eventual zero position for $\beta_c \rho$ is not included with this. Therefore, the limit conditions in the case of $E$-field modes are met for

$$\beta_c \cdot a = \frac{2\pi}{\lambda_{c,E}} a = j_{mn} . \qquad (1.195)$$

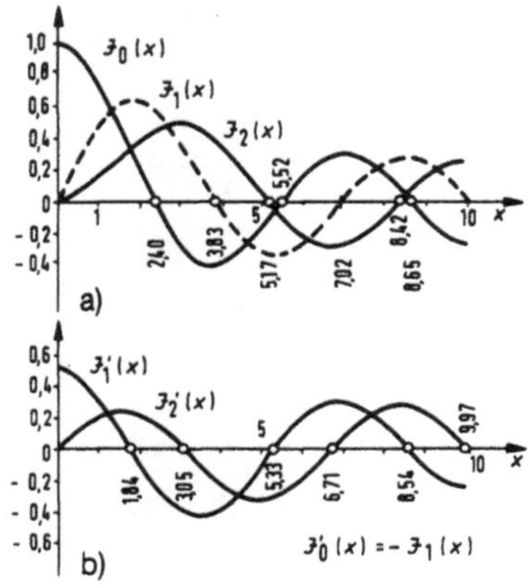


**Fig. 1.22** (a) Bessel functions and (b) derivations

Here the following relationship is valid, as with the rectangular hollow waveguide,

$$\beta_h^2 = \beta^2 - \beta_c^2 \, . \tag{1.196}$$

The critical wavelength $\lambda_{c,E}$ of the $E$-field mode in the circular hollow waveguide can be calculated as

$$\lambda_{c,E} = \frac{2\pi a}{j_{mn}} = \frac{\pi D}{j_{mn}} \quad \text{with} \quad \begin{array}{l} m = 0, 1, 2, \ldots \\ n = 1, 2, 3, \ldots \end{array} \tag{1.197}$$

Values for $\lambda_{c,E}/D$ are given in Table 1.1.

| Typ | $H_{11}$ | $E_{01}$ | $H_{21}$ | $E_{11}$ $H_{01}$ | $H_{31}$ | $E_{21}$ | $H_{41}$ | $H_{12}$ | $E_{02}$ | $E_{31}$ | $H_{51}$ | $H_{22}$ | $E_{12}$ $H_{02}$ |
|---|---|---|---|---|---|---|---|---|---|---|---|---|---|
| $\frac{\lambda_c}{D}$ | 1,706 | 1,306 | 1,029 | 0,820 | 0,748 | 0,612 | 0,591 | 0,589 | 0,569 | 0,492 | 0,490 | 0,469 | 0,448 |

The $E$-field modes in the circular hollow waveguide are characterized by the numbers $m$ and $n$ as subscripts:

*Subscript m* specifies (corresponding to the function cos $m\varphi$) the number of *nodal diameters*, at which the axial component $E_z$ of the electrical field (and $E_\rho, H_\varphi$) disappears; therefore, $m$ also represents the number of half-cycles in the field distributions along half the circumference.

*Subscript n* specifies (corresponding to the number of the zero position of $J_m$) the number of concentric *nodal circles* with the waveguide axis in a cross section, at which the axial electrical field intensity $E_z$ (and $E_\varphi, H_\rho$) disappears; therefore, $n$ also specifies the number of half-cycles in the field distribution along half the diameter.

Figure 1.23 illustrates this with the example of the $E_{22}$-mode for the magnetic lines of force, which with $E$-modes only run in cross-section planes.

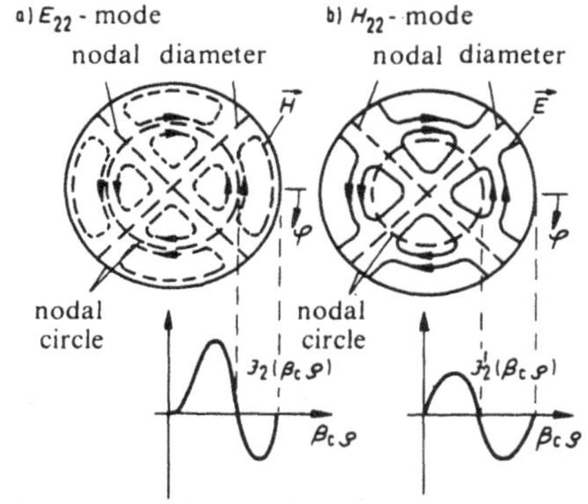


**Fig. 1.23** Nodal diameters and nodal circles for (a) the $E_{44}$-mode and (b) the $H_{22}$-mode

We will now explain $H$-waves with $E_z = 0$ in the circular hollow waveguide. For the field components, we find

$$\underline{H}_z = H_0 \cdot \cos(m\varphi) \cdot J_m(\beta_c \varrho) \cdot e^{j(\omega t - \beta_h \cdot z)} \tag{1.198}$$

$$\underline{H}_\varrho = H_0 \cdot \frac{\beta_h}{\beta_c} \cdot \cos(m\varphi) \cdot J_m'(\beta_c \varrho) \cdot e^{j(\omega t - \beta_h \cdot z + \pi/2)} \tag{1.199}$$

$$\underline{H}_\varphi = H_0 \cdot \frac{\beta_h}{\beta_c^2} \cdot \frac{m}{\varrho} \cdot \sin(m\varphi) \cdot J_m(\beta_c \varrho) \cdot e^{j(\omega t - \beta_h \cdot z - \pi/2)} \tag{1.200}$$

$$\underline{E}_z = 0 \tag{1.201}$$

$$\underline{E}_\varrho = \underline{H}_\varphi \cdot Z_{F,H} \tag{1.202}$$

$$\underline{E}_\varphi = -\underline{H}_\varrho \cdot Z_{F,H} \tag{1.203}$$

For $\beta_h = 2\pi/\lambda_h$ and $Z_{F,H}$ again the relationships of Eqs. (1.168) and (1.184) are valid as for the rectangular hollow waveguide. We obtain the critical wavelengths which are to be inserted there according to the following consideration from the limit condition $E_\varphi$, $H_\rho = 0$ for $\rho = a$. As the equations for the field components show, the zero positions of $J'_m(\beta_c\rho)$ now provide for adherance to these limit conditions. If we designate these zero positions with $j'_{mn}$, then we obtain

$$\beta_c \cdot a = \frac{2\pi}{\lambda_{c,H}} \cdot a = j'_{mn} \,. \tag{1.204}$$

It specifies once again $n$ as the number of the zero position, whereby an eventual zero position $\beta_c \cdot \rho$ is not included.

For the *critical wavelength* $\lambda_{c,H}$ of the field mode in a circular hollow waveguide, we obtain, therefore,

$$\lambda_{c,H} = \frac{2\pi a}{j'_{mn}} = \frac{\pi D}{j'_{mn}} \quad \text{with} \quad \begin{array}{l} m = 0, 1, 2, \ldots \\ n = 1, 2, 3, \ldots \end{array} \tag{1.205}$$

The numbers $m$ and $n$ serve as subscripts to indicate the field modes:

*Subscript m* specifies (corresponding to cos $m\,\varphi$) the number of *nodal diameters* at which the axial component $H_z$ of the magnetic field (and $H_\rho$, $E_\varphi$) disappears; therefore, $m$ also specifies the number of half-cycles in the field distributions along half the circumference.

*Subscript n* indicates (corresponding to the number of the zero position of $J'_m$) the number of concentric *nodal circles* with the waveguide axis in a cross section, at which the component $E_\varphi$, tangential to these circles, of the electrical field (and $H_\rho$) becomes zero; therefore, $n$ also specifies the number of half-cycles in the field distributions along half the diameter.

Figure 1.24 illustrates this with the example of the $H_{22}$-mode for the magnetic lines of force which run purely transverse in the case of $H$-modes.

In the circular hollow waveguide, the $E_{m0}$- and $H_{m0}$-field modes cannot exist, since in order to satisfy the limit conditions at the waveguide wall, a nodal circle must always appear there.

In Fig. 1.24, the path of electrical and magnetic lines of force and outlined for some field modes in the circular hollow waveguide.

E-waves possess only axial wall currents since the magnetic field is transversely directed.

In Table Ī.1 the critical wavelengths relative to the diameter, which result from the zero position of the Bessel functions or derivations in question, are shown for some $E$- and $H$-field modes in the circular hollow waveguide.

The $H_{11}$-field mode in the circular hollow waveguide with the longest limiting wavelength forms the primary or basic mode. Its field distribution and wall current distribution are similar to those of the $H_{10}$-mode in the rectangular hollow waveguide.

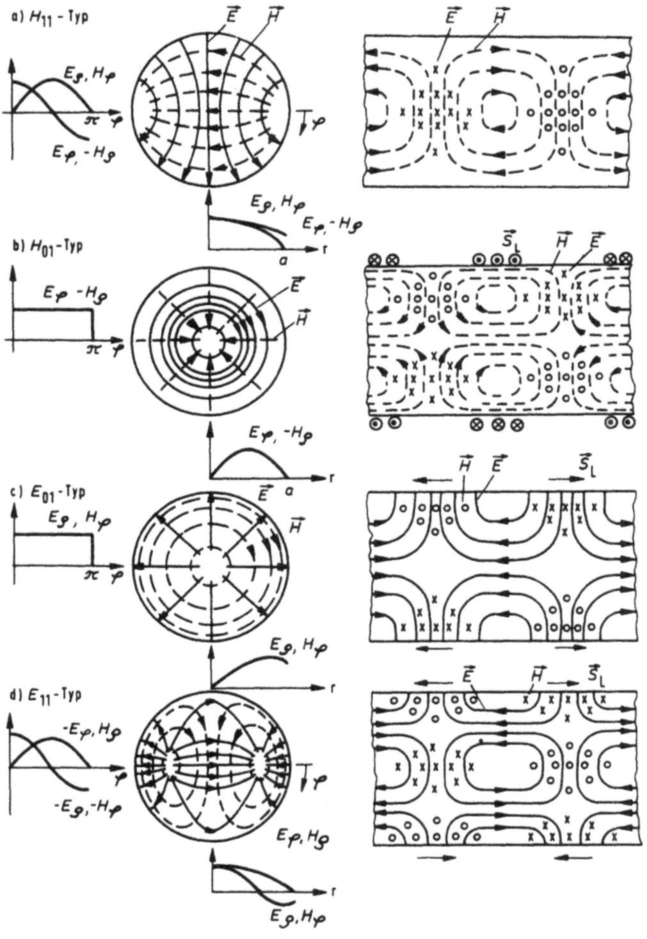


**Fig. 1.24** Field distribution in circular hollow waveguides

For the phase and group velocities in the circular hollow waveguide the same relationships of Eqs. (1.185) and (1.186) are valid as for the rectangular hollow waveguide, whereby the value of the circular hollow waveguide field mode in question is to be inserted for the critical wavelength $\lambda_c$.

The German standardization for circular hollow waveguides follows from DIN 47 302, part 2.

### 1.6.2.3 Wall Current Losses for the Hollow Waveguide

In the field observations employed above for the rectangular and circular hollow waveguides, the metal walls were assumed to be infinitely conductive. In practice, heat losses occur through wall currents as a result of the finite conductivity of the walls. Therefore, in addition to the reflection attenuation observed through focusing errors in the suppression range of the hollow waveguide ($f < f_c$), we also obtain an attenuation in the transmission range ($f > f_c$) through heat losses. In the case of good wall conductivity, the field will not change essentially with respect to the case of ideal conductivity. In order to approximate the attenuation, we proceed according to a procedure known as the *power loss method,* in which the field is used as a foundation in the lossless hollow waveguide. In this connection, the attenuation is calculated with the help of the active power $P_z$, which is transmitted in the $z$-direction in the hollow waveguide, and the power loss converted at the wall. We obtain $P_z$ through integration of the Poynting vector over the cross section of the hollow waveguide. The power loss (decrease in active power) per element of length $dP_z/dz$ can be calculated with the help of the surface resistance $R_d$ of the separation (Eq. (1.69)) and the tangentially directed magnetic field intensity $H_t$ which dominates there.

For the distributed attenuation resulting from ohmic losses, we obtain

$$\alpha_R = -\frac{1}{2}\, P_z \cdot \frac{dP_z}{dz} = \frac{R_d}{2} \cdot \frac{\oint H_t \cdot ds}{\int (\vec{E} x \vec{H}) \cdot d\vec{A}} \quad \text{with} \quad R_d = \sqrt{\frac{\pi \mu f}{\varkappa}}. \tag{1.206}$$

According to this procedure, we obtain for the distributed resistance attenuation losses:

a) for the $H_{10}$-wave in the rectangular hollow waveguide

$$\alpha_R = \sqrt{\frac{\pi \mu_0 c}{\varkappa}} \cdot \frac{2}{Z_0} \cdot \frac{\frac{a}{2b} + \left(\frac{\lambda_0}{2a}\right)^2}{a \cdot \sqrt{\lambda_0} \cdot \sqrt{1 - \left(\frac{\lambda_0}{2a}\right)^2}} = \frac{0{,}212}{\frac{a}{\mathrm{cm}} \cdot \sqrt{\frac{\lambda_0}{\mathrm{cm}}}} \cdot \frac{\frac{a}{2b} + \left(\frac{\lambda_0}{2a}\right)^2}{\sqrt{1 - \left(\frac{\lambda_0}{2a}\right)^2}} \frac{\mathrm{dB}}{\mathrm{m}}. \tag{1.207}$$

b) for the $H_{11}$-wave in the circular hollow waveguide

$$\alpha_R = \sqrt{\frac{\pi\mu_0 c}{\varkappa}} \cdot \frac{2}{Z_0} \cdot \frac{1}{D \cdot \sqrt{\lambda_0}} \cdot \frac{\left(\frac{\lambda_0}{1{,}706 \cdot D}\right)^2 + 0{,}4185}{\sqrt{1 - \left(\frac{\lambda_0}{1{,}706 \cdot D}\right)^2}} =$$

$$= \frac{0{,}212}{\frac{D}{\text{cm}} \cdot \sqrt{\frac{\lambda_0}{\text{cm}}}} \cdot \frac{\left(\frac{\lambda_0}{1{,}706 \cdot D}\right)^2 + 0{,}4185}{\sqrt{1 - \left(\frac{\lambda_0}{1{,}706 \cdot D}\right)^2}} \frac{\text{dB}}{\text{m}} . \tag{1.208}$$

c) for the $H_{01}$-wave in the circular hollow waveguide

$$\alpha_R = \sqrt{\frac{\pi\mu_0 c}{\varkappa}} \cdot \frac{2}{Z_0} \cdot \frac{1}{D \cdot \sqrt{\lambda_0}} \cdot \frac{\left(\frac{\lambda_0}{0{,}82 \cdot D}\right)^2}{\sqrt{1 - \left(\frac{\lambda_0}{0{,}82 \cdot D}\right)^2}} =$$

$$= \frac{0{,}212}{\frac{D}{\text{cm}} \cdot \sqrt{\frac{\lambda_0}{\text{cm}}}} \cdot \frac{\left(\frac{\lambda_0}{0{,}82 \cdot D}\right)^2}{\sqrt{1 - \left(\frac{\lambda_0}{0{,}82 \cdot D}\right)^2}} \frac{\text{dB}}{\text{m}} . \tag{1.209}$$

For the specific conductivity of the walls, the value for copper $\kappa = 56.18$ $(S \cdot \text{m/mm}^2)$ is used. With other conductive materials $\alpha = \alpha_{Cu} \cdot \sqrt{(\kappa_{Cu}/\kappa)}$ is used. Air is adopted as the dielectric of the hollow waveguide.

The attenuation formulas are valid for $f > 1.01 f_c$.

Because of the negligible depth of penetration $d$ in the microwave range, surface roughness leads to an increase in attenuation as a result of the extended current paths. If the peak-to-valley height is equal to the depth of penetration, then we obtain an increase in attenuation of approximately 60 percent. If it amounts to half the depth of penetration, then the increase is about 20 percent. In order to decrease the ohmic losses, we can polish or coat the inner wall surfaces. Since the wall currents only flow within a thin layer of the inner wall, a surface heat treatment of the metal separation (mostly copper, aluminum, or brass) with a well-conducting metal (silver or gold) suffices to reduce the surface resistance.

Humidity on the inner walls also increases the attenuation. Furthermore, the line attenuation increases in the presence of dielectric material in the hollow waveguide.

The occurrence of standing waves also increases the attenuation [by a factor of $2(1+s^2)/(1+s)^2$ with a constant generator power, and by $(1+s^2)/2s$ with a constant receiver power, $s$ = standing-wave ratio, Eq. (1.120)].

The attenuation caused by ohmic losses in the conductor wall decreases for increasing cross-section dimensions of the hollow waveguide.

The attenuation of the hollow waveguide, except for that of $H_{0n}$-waves in the circular hollow waveguide, decreases at first with an increasing frequency close to the critical frequency, passes quickly through a minimum, and then increases as a result of the growing losses caused by the skin effect. For $f \gg f_c$, then $\alpha \approx \sqrt{f}$ [compare with Eqs. (1.207), (1.208)]. In proximity to the critical frequency, the specified attenuation formulas with $a \to \infty$ for $f \to f_c$ become useless. In practice, the attenuation grows only slowly there, and the attenuation curve constantly crosses over into the curve of the reflection attenuation in the suppression range (Fig. 1.17).

A special feature of the attenuation behavior is shown by $H_{0n}$-waves in the circular hollow waveguide. With these waves, the magnetic field at the separation has only one component in the horizontal direction. The wall currents running parallel to it, therefore, possess only one component in the circumferential direction, and thus cannot cause any voltage drop in the horizontal direction. The magnetic horizontal field intensity decreases with an increasing frequency $f$ such as

$$\frac{1}{\sqrt{\left(\frac{f}{f_c}\right)^2 - 1}}$$

The transverse wall currents are thereby reduced and the attenuation decreases monotonically with the frequency (although the surface resistance increases), and does so for $f \gg f_c$ with $1/f^{3/2}$ [compare to Eq. (1.209)].

Of special significance in transmission over long distances (approximately $>1$ km) is the $H_{01}$-wave in the circular hollow waveguide because of the favorable behavior of the attenuation. What is unfavorable is that still other field modes are capable of propagation (compare with Table 1.1), and that they can be stimulated by inhomogeneities, such as contortions of the hollow waveguide. In this manner, power losses of the desired $H_{01}$-wave can be created by the transformation of wave modes, or signal distortions in the event that the other wave modes with changed phase velocities propagate and are transformed back again into $H_{01}$-waves. Therefore, the hollow waveguide must display a high attenuation for the undesired wave mode. What is suitable are hollow waveguides whose inner wall is formed by metal rings which are isolated from one another and are located next to each other, or by a wire coil of negligible size (on a dielectric intermediate layer) with loops separated from each other (helical hollow waveguide is also flexible), since these conduct the circular wall currents of the $H_{01}$-

wave well and the horizontal currents of the undesired wave mode poorly. The $E_{11}$-wave can be easily stimulated because it possesses the same phase velocity as the $H_{01}$-wave (compare with Table 1.1). The power transition at the $E_{11}$-wave can be kept negligible by taking measures that produce a difference in the phase velocities, as, for example, the mounting of a thin dielectric layer on the inner conductor wall.

By deviations from the ideal geometry (no exact circular cross section, shaft undulations), the attenuation of the $H_{01}$-wave again increases in practice at very high frequencies through a partial transformation of the $H_{01}$-wave into $H_{11}$- and $H_{12}$-waves. These are absorbed by a wave mode filter in order to avoid retransformation into the $H_{01}$-wave at centers of disturbance along the line which would lead to signal distortions; thereby, however, energy is withdrawn from the $H_{01}$-wave. We obtain, for example, attenuation values of approximately 1 dB/km with 70 mm hollow waveguides in the frequency range between 30 and 100 GHz. A transformation of 300,000 voice channels is possible on such a hollow waveguide. The information can be transmitted in the form of binary signals which are only slightly susceptible to interference.

### *1.6.2.4 Transfer Power*

The maximum peak transfer power with a high frequency circuit is restricted by the disruptive discharge which occurs with too high an electrical field intensity. The transfer power in a hollow waveguide is approximately proportional to the transverse plane of the hollow waveguide. It depends both on the wave mode and the frequency. As the dielectric strength in an air-filled hollow waveguide, we establish approximately 30 kV/cm (at $T = 20°C$, $P = 10$ N/cm² and dry air). As the upper limit of the flux density of the power, we obtain, therefore, with a non-reflecting termination at $f = 1.5\, f_c$ with the rectangular hollow waveguide for the $H_{10}$- and $H_{11}$-wave (the latter with $a = b$), $S_{max} = 445$ kW/cm²

$$\left(\frac{P_{max}}{A}\Big/\frac{\text{kW}}{\text{cm}^2} = 598 \cdot \sqrt{1 - \left(\frac{f_c}{f}\right)^2}\right)$$

and for the $E_{11}$-wave 500 kW/cm². For the circular hollow waveguide, we obtain with the $E_{01}$-wave $S_{max} = 540$ kW/cm², with the $H_{11}$-wave 425 kW/cm²

$$\left(\frac{P_{max}}{A}\Big/\frac{\text{kW}}{\text{cm}^2} = 568 \cdot \sqrt{1 - \left(\frac{f_c}{f}\right)^2}\right)$$

and with the $H_{01}$-wave 430 kW/cm² [1.3]. The transfer power is lessened by a factor $s$ in the case of a mismatched circuit and also by inhomogeneities in the circuit. We can usually maintain a safety factor of approximately 4 with respect to the calculated maximum value (irregularities in the cross section of the waveguide). The maximum transfer power can be increased by virtue of the fact that the circuit is operated with increased air pressure (or vacuum) and through the use of protective gas (for example, sulfide fluorine $SF_6$).

The permissible power densities of continuous-wave signals are restricted by the fact that the wall current losses must not lead to undue heating of the inner waveguide surfaces; these are essentially smaller than the corresponding permissible peak power densities.

### *1.6.2.5 Dielectric Material in Hollow Waveguides*

If a hollow waveguide is partially or completely filled with dielectric material, then the propagation constant, the oscillation resistance, and the critical wavelength change. In the case of a filling completely with low-loss dielectric material, the following is valid for the critical wavelength

$$\lambda_{c_\varepsilon} = \lambda_c \cdot \sqrt{\varepsilon_r} \tag{1.210}$$

$\lambda_c$ = un-filled critical wavelength, and, therefore, for the waveguide length

$$\lambda_{h_\varepsilon} = \frac{\dfrac{\lambda_0}{\sqrt{\varepsilon_r}}}{\sqrt{1 - \left(\dfrac{\lambda_0}{\lambda_{c_\varepsilon}}\right)^2}} = \frac{\lambda_0}{\sqrt{\varepsilon_r - \left(\dfrac{\lambda_0}{\lambda_c}\right)^2}} \tag{1.211}$$

The waveguide length and the phase velocity decrease with the introduction of dielectric material.

For the characteristic wave impedances, the following is true of *H*-waves

$$Z_{F,H_\varepsilon} = \frac{Z_0}{\sqrt{\varepsilon_r}} \cdot \frac{1}{\sqrt{1 - \left(\dfrac{\lambda_0}{\lambda_{c_\varepsilon}}\right)^2}} = \frac{Z_0}{\sqrt{\varepsilon_r - \left(\dfrac{\lambda_0}{\lambda_c}\right)^2}} < Z_{F,H} \tag{1.212}$$

and of *E*-waves

$$Z_{F,E_\varepsilon} = \frac{Z_0}{\sqrt{\varepsilon_r}} \cdot \sqrt{1 - \left(\frac{\lambda_0}{\lambda_{c_\varepsilon}}\right)^2} = Z_0 \cdot \sqrt{\frac{1 - \dfrac{1}{\varepsilon_r}\left(\dfrac{\lambda_0}{\lambda_c}\right)^2}{\varepsilon_r}}\,. \tag{1.213}$$

For $\lambda_0/\lambda_c > 1/\sqrt{2}$, then $Z_{F,E_\epsilon} > Z_{F,E}$.

If the hollow waveguide is filled with dielectric material, then the dielectric attenuation is, in general, essentially higher than the attenuation produced by wall current losses. In the case of small dielectric losses, the following is approximately valid for the distributed attenuation $\alpha_E$ as a result of dielectric losses

(with an arbitrary tube diameter)

$$\alpha_\varepsilon = \sqrt{\varepsilon_r} \cdot \frac{\pi}{\lambda_0} \cdot \tan \delta_\varepsilon \cdot \frac{1}{\sqrt{1 - \left(\frac{\lambda_0}{\lambda_{c_\varepsilon}}\right)^2}} \tag{1.214}$$

[compare with Eq. (1.93)].

If the hollow waveguide is only partially filled with dielectric material, then the strongest effect results when it is brought into the area of greatest electrical field intensity, therefore, for the $H_{10}$-wave in the center of the hollow waveguide. The bandwidth and the transfer power of the hollow waveguide can be increased in this manner. For $\lambda_n < \lambda_0$, a total reflection takes place at the transition between dielectric and air, and the majority of the energy is transferred into the dielectric material.

The introduction of dielectric material into hollow waveguides for the purpose of a phase shift will be dealt with in Chapter 3 on circuit components.

### *1.6.2.6 Ridge Hollow Waveguide*

With a continuously ridged constriction of the cross-section area of the hollow waveguide, according to Fig. 1.25 (also constructed unilaterally or semicircularly because of a better disruptive strength), which functions as a condensor load (see also capacitative inserts), we can increase the critical wavelength of the $H_{10}$-wave (diagrams by S. Hopper in *Trans. I.R.E.*, MTT-3, 1955, no. 5) without fundamentally changing the critical wavelength of the $H_{20}$-wave. In this manner, the frequency bandwidth of the hollow waveguide can be increased. The transfer power decreases as a result of the decrease in the cross-section area.

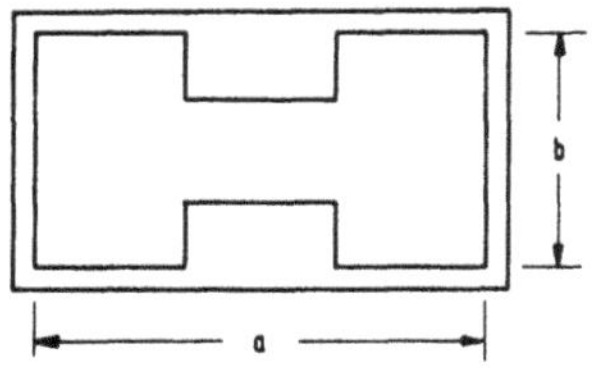


**Fig. 1.25** Ridge hollow waveguide

### *1.6.2.7 Flexible Hollow Waveguides*

Hollow waveguides are also manufactured which are flexible within certain limits. The flexibility can be attained by the fact that the waveguide separation is formed by twisted spiral parts which interlock, that it is made without joints from a corrugated resilient metal (corrugated tube), or that it is made from components which are mechanically separated from each other and which are electrically connected by the short circuits transformed in the separation (see the choke connector). The waveguide is encased in a rubber coating.

### *1.6.2.8 Higher Field Modes in L-Waveguides*

In $L$-waveguides, such as the coaxial transmission line, the stripline, and the two-wire circuit, $E$- and $H$-waves can occur in addition to the TEM-mode at higher frequencies.

In the coaxial transmission line, the $H_{11}$-wave possesses the largest critical wavelength with

$$\lambda_c = 2{,}95\,(r_a + r_i) \cdot \sqrt{\varepsilon_r}\,. \tag{1.215}$$

The pattern of the lines of force for this field mode resembles that of the $H_{11}$-mode in the circular hollow waveguide. In the vicinity of the inner conductor, the electrical lines of force are approximately directed upon it and the magnetic lines of force are directed away from it.

We generally use a coaxial transmission line below the critical frequency of the $H_{11}$-wave in order to maintain a pure TEM-mode. At inhomogeneities of the circuit, then, solely aperiodic evanescent $E$- and $H$-interference fields can form. If the excitation and the interference patterns of the circuit are exactly cylindrically symmetrical, then only cylindrically symmetric interference fields can occur. In this case, then, the main interference field is in the $E_{01}$-field, which has, however, an essentially shorter critical wavelength

$$\lambda_c = 2{,}029\,(r_a - r_i) \cdot \sqrt{\varepsilon_r} \tag{1.216}$$

and dies down very quickly once it is away from the center of disturbance.

In the frequency range $f < 1$ GHz, we mostly use coaxial transmission lines for high frequency energy transmission because the cross-section dimensions of hollow waveguides become too cumbersome. For frequencies of approximately $f > 20$ GHz, coaxial cables become unsuitable because the small dimensions necessary in order to avoid higher wave forms cause too many losses.

### *1.6.2.9 Application of Line Theory to Hollow Waveguides*

As was pointed out in sec. 1.5.1, tension, current, and, therefore, impedances in the cross-section plane of TEM-wave circuits can be clearly defined. The behavior of the amplitude and phase of a TEM-wave's field components along a circuit corresponds to that of the transverse components of $E$- or $H$-waves. It is, therefore, obvious that we must introduce corresponding definitions for circuits which also conduct $E$- or $H$-waves.

We will, therefore, consider a charge transported in the positive direction through the cross-section surface $A$ at point $z$.

With TEM-waves the following is valid

$$P(z) = \frac{1}{2} \operatorname{Re}\{\underline{U}(z) \cdot \underline{I}^*(z)\} = \frac{1}{2}|\underline{I}|^2 \cdot \operatorname{Re}\{\underline{Z}(z)\} = \frac{1}{2}|\underline{U}|^2 \cdot \operatorname{Re}\{\underline{Z}(z)\} \quad (1.217)$$

with the line impedance

$$\underline{Z}(z) = \frac{\underline{U}(z)}{\underline{I}(z)} . \quad (1.218)$$

With $E$- or $H$-waves, we obtain with the help of the Poynting vector

$$\underline{\vec{P}}' = \frac{1}{2} \underline{\vec{E}} \times \underline{\vec{H}}^* = \frac{1}{2} \underline{E}_t \cdot \underline{H}_t^* \cdot \vec{e}_z \quad (1.219)$$

for the transported charge

$$P(z) = \frac{1}{2} \operatorname{Re} \left\{ \int_{(A)} \underline{E}_t(x, y) \cdot \underline{H}_t^*(x, y) \cdot dA \right\} . \quad (1.220)$$

Here $\underline{E}_t$ and $\underline{H}_t$ are the complex vectors of the transverse field components which are perpendicular to one another in space. These can be defined by comparing Eqs. (1.217) and (1.220).

$$\operatorname{Re}\{\underline{U}(z) \cdot \underline{I}^*(z)\} = \operatorname{Re} \left\{ \int_{(A)} \underline{E}_t(x, y) \cdot \underline{H}_t^*(x, y) \cdot dA \right\} . \quad (1.221)$$

If we write for the transverse field components at point $z$

$$\underline{E}_t(x, y)|_z = \underline{U}(z) \cdot f(x, y) \quad (1.222)$$

and

$$\underline{H}_t(x, y)|_z = \underline{I}(z) \cdot g(x, y) \quad (1.223)$$

then Eq. (1.221) will be satisfied if

$$\int_{(A)} f(x, y) \cdot g(x, y) \cdot dA = 1 . \quad (1.224)$$

The definition according to Eqs. (1.222) and (1.223) for equivalent voltage and current is not clear, since a combination of $\underline{\tilde{U}}, \underline{\tilde{I}}$, with $\underline{\tilde{U}} = \underline{U}/c$ and $\underline{\tilde{I}} = c \cdot \underline{I}$, $c$ is a constant, leads to the same charge $\underline{\tilde{U}} \cdot \underline{\tilde{I}}^* = \underline{U} \cdot \underline{I}^* = \underline{U} \cdot \underline{I}$. $\underline{Z}(z) = \underline{U}(z)/\underline{I}(z)$ is equally unclear, since $\underline{\tilde{Z}}(z) = \underline{\tilde{U}}(z)/\underline{\tilde{I}}(z) = \underline{Z}(z)/c^2$. We usually choose the characteristic wave impedance of the hollow waveguide for the $E$- or $H$-wave in question for $\underline{Z}(z)$. Then,

$$\frac{\underline{U}(z)}{\underline{I}(z)} = \frac{\underline{E}_t(z)}{\underline{H}_t(z)} . \quad (1.225)$$

The transverse field components at point z can be combined with those of an incident or reflected wave

$$\underline{E}_t(z) = \underline{E}_{t,h}(z) + \underline{E}_{t,r}(z) \tag{1.226}$$

and

$$\underline{H}_t(z) = \underline{H}_{t,h}(z) + \underline{H}_{t,r}(z) = \underline{Z}_F \cdot [\underline{E}_{t,h}(z) - \underline{E}_{t,r}(z)] \,. \tag{1.227}$$

Analogous to the definition of reflectivity at point $z$ of incident or reflected voltage or current waves [Eq. (1.100)], we obtain the field approach

$$\underline{r}(z) = \frac{\underline{E}_{t,r}(z)}{\underline{E}_{t,h}(z)} = -\frac{\underline{H}_{t,r}(z)}{\underline{H}_{t,h}(z)} \,. \tag{1.228}$$

We obtain with Eqs. (1.225) to (1.228)

$$\frac{\underline{U}(z)}{\underline{I}(z)} = \underline{Z}_F \cdot \frac{1 + \underline{r}(z)}{1 - \underline{r}(z)} \,. \tag{1.229}$$

The applicability of the impedance concept and the reflectivity constant for a specific cross section in a hollow waveguide is valid, at times, only for a specific wave mode with a clear polarization. The cross section of the waveguide examined is to be placed far enough away from interference patterns so that the disturbing fields die down. Then we can introduce relative resistance or conductance for hollow waveguide networks and calculate with equivalent networks. We can also use circuit diagrams as, in the case of TEM-waveguides, since the transformation of impedance and reflectivity takes place in the same manner [Eqs. (1.102 and (1.103)] and the reflectivity $\underline{r}(z)$ along the circuit also changes in the same manner [Eq. (1.109)].

## 1.7 STRIPLINES

In the case of various microstrip lines, we are dealing with circuit forms derived from the stripline. These are especially suited for the construction of microwave circuits (see also Chapter 4). In Fig. 1.24, several designs of striplines with lines of force for the basic field mode are represented. Because of its simple structure, we use the *microstrip line* quite often (Fig. 1.26b), in which a layered conductor is mounted on a conducting carrier plate with an intermediate dielectric layer. In order to attain better shielding, a second specularly symmetrical configuration is stratified (sandwich design), and we obtain the symmetric or *triplate-stripline* (Fig. 1.26c), or the microstrip line is put into a conducting case. In the case of the designs shown in Fig. 1.26d (High-Q-triplate) and 1.26e, the use of dielectric material in order to reduce the losses is greatly restricted. Here it is particularly favorable (Fig. 1.26) that the dielectric material be in an area of relatively low field intensity. The *chamber line* in Fig. 1.26f also possesses relatively low attenuation. The *slot-line* and the *coplanar line* [double slit-line (Fig. 1.26d, e)] offer the advantage that structural elements connected in parallel can be built in without drilling through the carrier plate. Better, higher oscillation resistances (> 100Ω) can be realized with slot-lines than with microstrip lines, since the latter show high losses there. Otherwise, they are well suited to the construction of ferrite components because of their horizontal magnetic field intensity (Chapter 5). Additional stripline forms are represented in Fig. 4.13.

In the case of a homogeneous circuit with more than one conductor and a uniform dielectric, the basic field mode is a TEM-mode. Since with striplines a small portion of the lines of force also runs partially through air (different phase velocity) forming a leakage field from the edge of the conductor path outward, these waves also possess negligible field components in the direction of propagation ($E_z, H_z, EH$-waves).

With slot-lines the basic field mode is, initially, an $H$-mode, a so-called quasi-$H$-mode; the horizontal components of $E$ remain small. With other striplines, the horizontal components of $E$ and $H$ are to be disregarded for relatively low frequencies, and for which the dispersion remains negligible; thus, as a basic mode we can adopt an approximate TEM-mode, a so-called *quasi-TEM-mode.* For the phase velocity, the TEM-mode wavelength and the oscillation resistance of a TEM-wave, Eqs. (1.91), (1.90), and (1.92) are valid. For the calculation of the characteristic line magnitudes of the approximately assumed TEM-mode, the applied procedures are the conformal representation (used by H.A. Wheeler), the variation method, the jointing plane method, or the Fourier-integral method with the Green function (calculation with the help of the so-called "dielectric" Green functions according to T.G. Bryant and J.A. Weiss) with the corresponding numerical evaluations.

For the *microstrip line,* we obtain (according to H.A. Wheeler, Hammerstadt)

$$Z_L = \frac{Z_0}{\sqrt{\varepsilon_{eff}}}$$

with

$$Z_0 \approx 60\ \Omega \cdot \ln\left(\frac{8h}{b} + \frac{b}{4h}\right) \qquad \text{for} \quad \frac{b}{h} \leqq 1 \qquad (1.230)$$

and

$$Z_0 \approx \frac{120\ \pi\Omega}{1{,}393 + \frac{b}{h} + 0{,}667 \cdot \ln\left(1{,}44 + \frac{b}{h}\right)} \qquad \text{for} \quad \frac{b}{h} \geqq 1$$

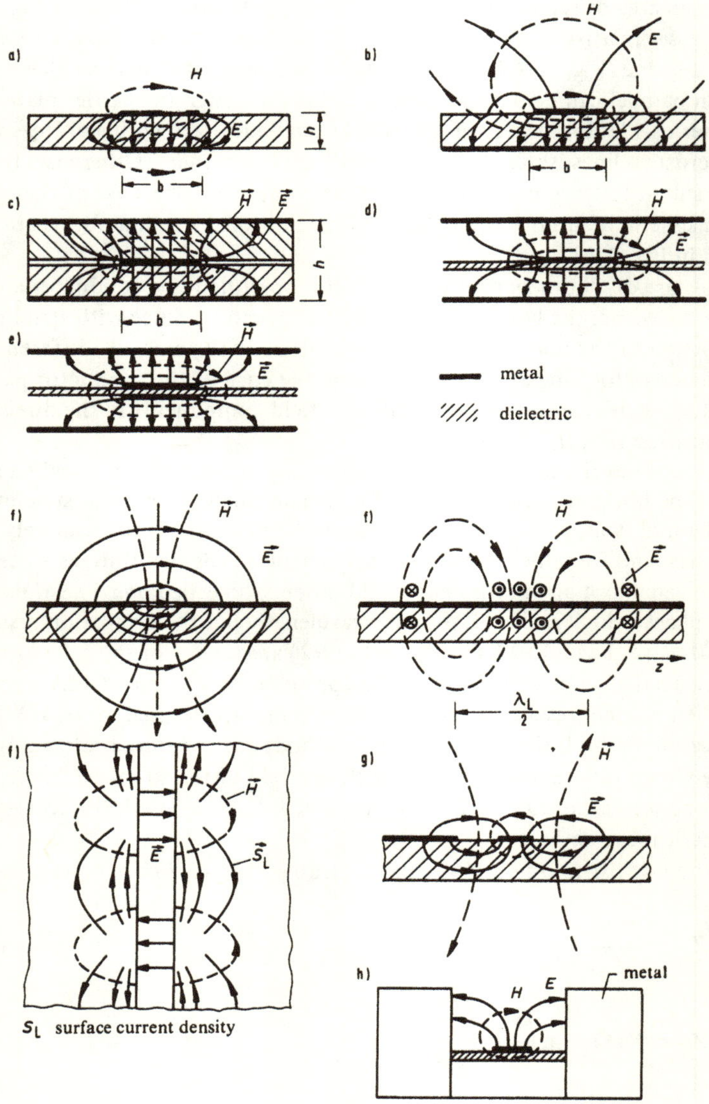


**Fig. 1.26** Striplines: (a) stripline, (b) microstrip line, (c) symmetrical stripline (triplate line), (d, e) High-Q triplate line, (f) slot-line, (g) double slot-line or coplanar line, (h) chamber line

and the effective inductivity

$$\varepsilon_{\text{eff}} \approx \frac{\varepsilon_r + 1}{2} + \frac{\varepsilon_r - 1}{2}\left[\frac{1}{\sqrt{1 + 12\dfrac{h}{b}}} + 0.04\left(1 - \frac{b}{h}\right)^2\right] \quad \text{for} \quad \frac{b}{h} \leqq 1 ,$$

$$\varepsilon_{\text{eff}} \approx \frac{\varepsilon_r + 1}{2} + \frac{\varepsilon_r - 1}{2} \cdot \frac{1}{\sqrt{1 + \dfrac{10h}{b}}} \quad \text{for} \quad \frac{b}{h} \geqq 1 . \tag{1.231}$$

Here $b$ is the width of the strip and $h$ the thickness of the carrier plate. Because of the stratified dielectric substrate-air, the effective inductivity is $\epsilon_{\text{eff}} \neq \epsilon_r$. For the TEM-mode wavelength, we obtain

$$\lambda_L = \frac{\lambda_0}{\sqrt{\varepsilon_{\text{eff}}}} . \tag{1.232}$$

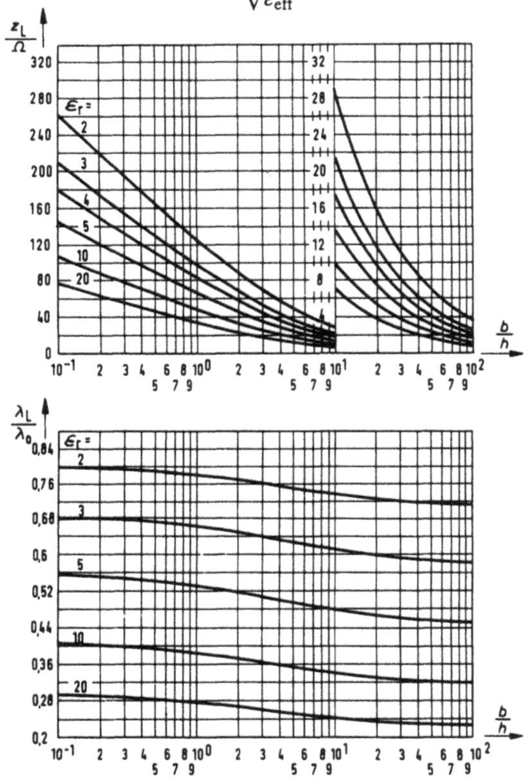


**Fig. 1.27** TEM-mode (a) oscillation resistance and (b) velocity rate of the microstrip line with $\epsilon_r$ as the parameter

Figure 1.27 represents, for the microstrip line (Fig. 1.26b), the path of the TEM-mode oscillation resistance $Z_L$ and the velocity rate $\lambda_L/\lambda_0 = 1/\sqrt{(\epsilon_{eff})}$ as a function of the relationship between the bandwidth $b$ and the thickness $h$ of the carrier plate for some values of $\epsilon_r$.

In order to avoid higher field modes, we must choose

$$b < \frac{\lambda_L}{2}, \quad h < \frac{\lambda_L}{4} \tag{1.233}$$

(compare with Fig. 1.26).

The expansion of the leakage fields decreases for larger $\epsilon_r$ values of the column support, which is called the substrate. Higher $\epsilon_r$ values make a corresponding reduction of the dimensions possible as a result of the reduction in the TEM-mode wavelength.

For the oscillation resistance of the symmetrical stripline (triplate-line), the following is approximately valid

$$Z_L = \frac{60\ \Omega}{\sqrt{\varepsilon_r}} \cdot \ln\left(2 \coth \frac{\pi b}{4h}\right) \quad \text{for} \quad Z_L > \frac{30\ \pi}{\sqrt{\varepsilon_r}}\ \Omega$$

and (1.234)

$$Z_L = \frac{15\ \pi^2}{\sqrt{\varepsilon_r} \cdot \ln(2 \cdot e^{\pi b/2\,h})} \quad \text{for} \quad Z_L < \frac{30\ \pi}{\sqrt{\varepsilon_r}}$$

with $h$ = substrate height
$b$ = distance of the carrier plates

The effective inductivity here is $\epsilon_{eff} \approx \epsilon_r$. The formulas valid for a conductor thickness → 0 can be used with good approximation for the usual small thickness of the layer ($< 70\ \mu m$).

The column support should be low-loss in the interests of a low attenuation. It should exhibit a uniform thickness, composition, and a smooth surface; possess a low thermal expansion; be resistant to mechanical strain, chemicals, and influences of temperature; and be water-repellent and well machined. Here a suitable compromise must be chosen.

The following are used as substrates (see also Table 4.1):

*Synthetic materials:* (reinforced glass fiber) Teflon, Polystyrene, Polyolefine ($\epsilon_r \approx 2.5$; $\tan\delta \approx 5 \cdot 10^{-4}$),

*Ceramic materials:* $Al_2O_3$ ($\epsilon_r \approx 9.7$; $\tan\delta \approx 3 \cdot 10^{-4}$ with 99.5 percent material), $Al_2O_3$ such as sapphire ($\epsilon_r \approx 10$-$12$; $\tan\delta < 10^{-4}$, BeO ($\epsilon_r \approx 6.4$; $\tan\delta \approx 3 \cdot 10^{-4}$ with 99.5 percent material, high heat conductivity!), quartz ($\epsilon_r \approx 3.75$; $\tan\delta < 10^{-4}$ and materials with very high inductivity such as $TiO_2$ ($\epsilon_r \approx 85$; $\tan\delta \approx 4 \cdot 10^{-3}$).

The specified values refer to 10 GHz and 25°C (see also Table 4.1).

A small substrate plate thickness of 0.635 mm (= 25 mil = $25 \cdot 10^{-3}$ in) is customary.

For coating the carrier plate (layer thickness up to approximately 70 $\mu$m), we use a well conducting metal such as copper, silver, or gold. In order to improve the adhesion, we may attach a thin, highly resistant metallic intermediate layer (e.g., Cr, Ti, Ta) between the substrate and the conductor.

The attenuation of striplines is composed of the attenuation components of the dielectric, the conductor, and the effects of radiation. Assuming that the field only runs in dielectric material, the distributed attenuation caused by dielectric losses can be approximately calculated according to Eq. (1.93) (losses from microstrip lines $\approx$ 1/substrate thickness). The attenuation through ohmic losses in the conductors climbs proportionally with $\sqrt{f}$. The attenuation of striplines (order of magnitude of 0.01 -0.1 dB/cm) is usually higher than the corresponding hollow waveguides. Striplines are, therefore, suited for the realization of high quality resonators and for long distance transmission. An additional disadvantage is the low transfer power. Also favorable by comparison with the hollow waveguide, in addition to its simpler construction, is the larger bandwidth resulting from the smaller dispersion.

In stripline technology the construction of space and weight saving microwave circuits is possible, which can be relatively cheaply and easily manufactured and reproduced. They are, therefore, suited for mass production. In this method the circuits are mounted on the carrier plate in a similar fashion to the well known printed circuit board. The circuit components can then be integrated or installed into the so-called hybrid technology as small (as compared to the wavelengths) additional concentrated structural elements. Stripline engineering is applied in the frequency range of approximately 1 to 35 (100) GHz. At high frequencies the realization of the necessary small dimensions [Eq. (1.134)] becomes difficult.

## 1.8 SURFACE WAVES

An emanation of electromagnetic waves is also possible along the boundary surfaces of two media with different material constants (and, therefore, different phase velocities) in the form of so-called surface waves (for example, along the ground plane).

In the following section, we will first investigate the simple case of a plane, infinitely long boundary surface for a medium (1) with the material constants $\epsilon_1, \mu_1, \kappa_1$ and a medium (2) with $\epsilon_2, \mu_2, \kappa_2$.

### 1.8.1 Plane Boundary Surface

The junction plane between the two media is the $xz$-plane (Fig. 1.28 and Fig. 1.29). Medium (1) is air with

$$\varepsilon_{r_1} = \mu_{r_1} = 1 \quad \text{and} \quad \varkappa_1 = 0 \,. \tag{1.235}$$

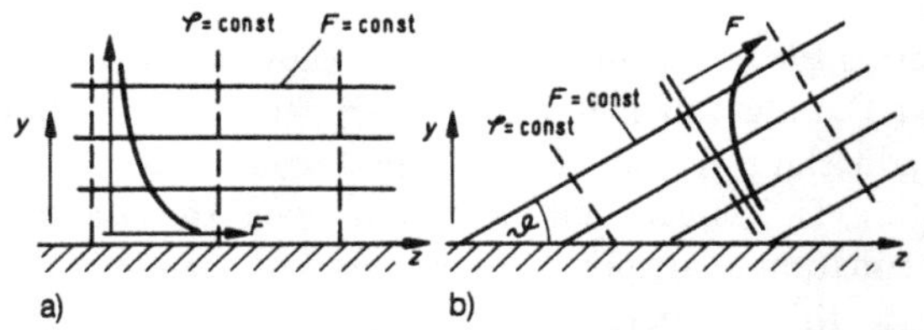


**Fig. 1.28** Amplitude and phase of (a) a fixed wave and (b) an emissive inhomogeneous wave

Because of the assumed infinitely long ground plane, the field magnitudes $F$ ($E$, or, respectively, $H$) are independent of $x$. With $\partial/\partial x = 0$, Maxwell's equations yield the wave equation [compare with (1.157)] :

$$\frac{\partial^2 \vec{F}}{\partial y^2} + \frac{\partial^2 \vec{F}}{\partial z^2} - \gamma^2 \vec{F} = 0 \tag{1.236}$$

with the expression known from the plane wave [Eq. (1.30)] :

$$\gamma^2 = \mathrm{j}\omega\mu\varkappa - \omega^2\mu\varepsilon = \mathrm{j}\omega\mu\,(\varkappa + \mathrm{j}\omega\varepsilon) \,. \tag{1.237}$$

If we insert the statement for the field magnitudes

$$\underline{F} = \underline{F}_0 \cdot \mathrm{e}^{\gamma_y \cdot y - \gamma_z \cdot z} \tag{1.238}$$

into the differential equation, then we obtain the condition

$$\gamma_y^2 + \gamma_z^2 = \gamma^2 \,. \tag{1.239}$$

We will first examine the relationships in medium (1) air ($y \geqslant 0$) with

$$\gamma^2 = -\beta_0^2 = -\left(\frac{2\pi}{\lambda_0}\right)^2 = -\left(\frac{\omega}{c}\right)^2 = -\,\omega^2\mu_0\varepsilon_0 \,. \tag{1.240}$$

with

$$\gamma_y = \alpha_y + \mathrm{j}\beta_y \tag{1.241}$$

and

$$\gamma_z = \alpha_z + \mathrm{j}\beta_z \tag{1.242}$$

Equation (1.239) yields

$$\gamma^2 = \alpha_y^2 - \beta_y^2 + \alpha_z^2 - \beta_z^2 + j(2\alpha_y \cdot \beta_y + 2\alpha_z \cdot \beta_z) . \qquad (1.243)$$

Since $\gamma^2$ is real, we obtain the following two conditions

$$\gamma^2 = \alpha_y^2 + \alpha_z^2 - \beta_y^2 - \beta_z^2 \qquad (1.244)$$

and

$$\alpha_y \cdot \beta_y + \alpha_z \cdot \beta_z = 0 . \qquad (1.245)$$

We assume wave propagation in the $z$-direction. Other cases can be discussed:

a) $\alpha_y = 0$; this means constant field amplitudes in the $y$-direction. With $\beta_z < 0$ (propagation in the $z$-direction), $\alpha_z = 0$ follows from Eq. (1.245). The wave is, therefore, also unattenuated in the $z$-direction. We are dealing with a homogeneous plane wave that propagates in the $z$-direction.

b) $\alpha_z = 0$ and $\alpha_y \neq 0$: $\beta_y = 0$ follows from Eq. (1.245) and $\beta_z^2 = \alpha_z^2 - \gamma^2 = \alpha_y^2 + (\omega/c)^2$ follows from Eq. (1.244). Since $\beta_z = \omega/v_p$, this means that the phase velocity of the wave is less than the speed of light:

$$v_p < c . \qquad (1.246)$$

The local variation of the field magnitudes according to Eq. (1.238) is now obtained through complex notation as

$$\underline{F} = \underline{F}_0 \cdot e^{\alpha_y \cdot y} \cdot e^{-j\beta_z \cdot z} . \qquad (1.247)$$

For constant amplitude surfaces, we obtain $y$ = const, therefore, planes parallel to the boundary plane. The planes of constant phase $z$ = const, are perpendicular to them. In the $y$-direction the field amplitudes change exponentially. These relationships are represented in Fig. 1.28a. A cross-attenuated wave such as this, which propagates along the boundary plane with a phase velocity $v_p < c$, is designated as a *fixed wave* or a *surface wave.* Whereas in the case of homogeneous waves in free space, or in circuits, constant phase surfaces are also constant amplitude surfaces, the phase and amplitude surfaces of waves that also designated as inhomogeneous waves do not coincide.

c) $\alpha_z < 0, \beta_y < 0$. The wave propagating in the $z$-direction moves toward the boundary surface ($\beta_y < 0$). Following from Eq. (1.245)

$$\alpha_y = -\frac{\alpha_z \cdot \beta_z}{\beta_y} < 0 ; \qquad (1.248)$$

the field amplitudes decrease exponentially in the $y$-direction. We obtain the constant amplitude surfaces of the wave by setting the exponents of the e-function equal to a constant [Eq. (1.238)] with a real argument: we obtain $\alpha_y \cdot y - \alpha_z = c_1$, or $y = \alpha_z/\alpha_y \cdot z - c_1$; and with Eq. (1.248): $y = -\beta_y/\beta_z \cdot z - c_1$. The planes of constant oscillation phase can be obtained by setting the exponent of the e-function constant with an imaginary argument. These planes are described by $\beta_y \cdot y - \beta_z \cdot z = c_2$, or $y = \beta_z/\beta_y \cdot z - c_2$. The planes of constant amplitude are inclined at the angle

$$\vartheta = \arctan\left(-\frac{\beta_y}{\beta_z}\right) \tag{1.249}$$

with respect to the boundary plane, and the planes of constant phase are perpendicular to them.

d) $\alpha_z > 0, \beta_y < 0$: the wave propagating in the $z$-direction moves away from the boundary plane ($\beta_y < 0$). Now, $\alpha_y = -(\alpha_z \cdot \beta_z)/\beta_y > 0$. This means that the wave field amplitudes in the $y$-direction increase exponentially. The planes of constant amplitude are inclined at the angle

$$\vartheta = \arctan \frac{\beta_z}{\beta_y} \tag{1.250}$$

with respect to the junction plane, and the planes of constant phase are perpendicular to them (Fig. 1.28b).

The phase velocity of the wave along the boundary plane is greater than the speed of light:

$$v_p > c\,. \tag{1.251}$$

Next, we describe a radiation of energy in a direction perpendicular to the phase fronts at an angle ($\vartheta$). This inhomogeneous wave, for which the phase velocity along the boundary plane is greater than the speed of light and a reflection takes place, is called a *radiation wave* or *leaky wave.* The exponential increase in amplitude toward the outside demonstrates that pure radiating waves do not exist. They describe the emanation of so-called radiating antennas. These antennas emit continuously along the entire boundary structure.

Surface waves are *bound* to the boundary structure. A reflection only takes place at discontinuities of the surface wave circuit, such as contortions, at the beginning and end of the circuit, or through a neutralization of the energy by coupled emitters.

In the following, we will more closely investigate an example of a bound inhomogeneous wave.

Of practical interest is the low attenuation transverse-magnetic wave ($E_{00}$-

wave), also called the Zenneck-wave after J. Zenneck who theoretically investigated it in 1907.

The direction of propagation of this wave is the $z$-direction. It possesses the field components $H_x$, $E_y$, and $E_z$. With the help of Maxwell's equations, we obtain for the field components satisfying the wave equation (1.236), media (1) and (2), the solutions

$$\underline{E}_{z_1} = \underline{A}_1 \cdot e^{\gamma_y^{(1)} \cdot y} \cdot e^{-\gamma_z \cdot z} \tag{1.252}$$

$$\underline{E}_{y_1} = \frac{\gamma_z}{\gamma_y^{(1)}} \underline{A}_1 \cdot e^{\gamma_y^{(1)} \cdot y} \cdot e^{-\gamma_z \cdot z} \tag{1.253}$$

$$\underline{H}_{x_1} = - \frac{\gamma^{(1)}}{\gamma_y^{(1)}} \cdot \sqrt{\frac{\varkappa_1 + j\omega\varepsilon_1}{j\omega\mu_1}} \underline{A}_1 \cdot e^{\gamma_y^{(1)} \cdot y - \gamma_z \cdot z} \tag{1.254}$$

$$\underline{E}_{z_2} = \underline{A}_2 \cdot e^{\gamma^{(2)} \cdot y} \cdot e^{-\gamma_z \cdot z} \tag{1.255}$$

$$\underline{E}_{y_2} = \frac{\gamma_z}{e^{\gamma_y^{(2)}}} \cdot \underline{A}_2 \cdot e^{\gamma_y^{(2)} \cdot y} \cdot e^{-\gamma_z \cdot z} \tag{1.256}$$

$$\underline{H}_{x_2} = - \frac{\gamma^{(2)}}{e^{\gamma_y^{(2)}}} \cdot \sqrt{\frac{\varkappa_2 + j\omega\varepsilon_2}{j\omega\mu_2}} \cdot \underline{A}_2 \cdot e^{\gamma_y^{(2)} \cdot y} \cdot e^{-\gamma_z \cdot z} \tag{1.257}$$

Here the material constants for media (1) and (2), respectively, are to be inserted for $\gamma^{(1)}$ and $\gamma^{(2)}$ in Eq. (1.237).

The field magnitudes must meet the limit conditions at the junction plane $y = 0$: following from the stability of the tangential component $E_z$ at the junction plane $\underline{E}_{z_1}$ $(y = 0) = \underline{E}_{z_2}$ $(y = 0)$, we have

$$\underline{A}_1 = \underline{A}_2 = \underline{E}_{z_0} \tag{1.258}$$

The stability of the tangential component $H_x$ for $y = 0$ yields, if we set $\mu_1 = \mu_2 = \mu_0$, in order to simplify,

$$\frac{\gamma_y^{(2)}}{\gamma_y^{(1)}} = \frac{\gamma^{(2)}}{\gamma^{(1)}} \cdot \sqrt{\frac{\varkappa_2 + j\omega\varepsilon_2}{\varkappa_1 + j\omega\varepsilon_1}} . \tag{1.259}$$

If the medium (1) is composed of air, then we obtain with

$$\gamma^{(1)} = j\omega \sqrt{\mu_0 \, \varepsilon_0} \tag{1.260}$$

the relation

$$\frac{\gamma_y^{(2)}}{\gamma_y^{(1)}} = - \frac{\varkappa_2 + j\omega\varepsilon_2}{j\omega\varepsilon_0} = - \left(\frac{\gamma^{(2)}}{\gamma^{(1)}}\right)^2 . \tag{1.261}$$

With Eq. (1.239), we obtain for the media (1) and (2)

$$(\gamma_y^{(1)})^2 + \gamma_z^2 = (\gamma^{(1)})^2 \tag{1.262}$$

$$(\gamma_y^{(2)})^2 + \gamma_z^2 = (\gamma^{(2)})^2 \,. \tag{1.263}$$

From this we obtain

$$\gamma_z^2 = \frac{(\gamma^{(1)})^2 \cdot (\gamma^{(2)})^2}{(\gamma^{(1)})^2 + (\gamma^{(2)})^2} \tag{1.264}$$

and

$$(\gamma_y^{(1)})^2 = \frac{(\gamma^{(1)})^4}{(\gamma^{(1)})^2 + (\gamma^{(2)})^2} \tag{1.265}$$

$$(\gamma_y^{(2)})^2 = \frac{(\gamma^{(2)})^4}{(\gamma^{(1)})^2 + (\gamma^{(2)})^2} \,. \tag{1.266}$$

If the medium (2) is a good conductor with

$$\varkappa_2 \gg \omega\varepsilon_2 \tag{1.267}$$

then we obtain with

$$(\gamma^{(1)})^2 = -\omega^2 \mu_0 \varepsilon_0 \tag{1.268}$$

and

$$(\gamma^{(2)})^2 = \mathrm{j}\omega\mu_0\varkappa_2 \tag{1.269}$$

the propagation constant $\gamma_z$ in the $z$-direction

$$\gamma_z = \mathrm{j}\,\frac{\omega}{c} \cdot \sqrt{\frac{1}{1 + \mathrm{j}\,\dfrac{\omega\varepsilon_0}{\varkappa_2}}} \approx \frac{\omega^2 \cdot \varepsilon_0}{2c \cdot \varkappa_2} + \mathrm{j}\,\frac{\omega}{c} \tag{1.270}$$

and the propagation constants in the $y$-direction for both media

$$\gamma_y^{(1)} = -\alpha_y^{(1)} + \mathrm{j}\beta_y^{(1)} \approx -\frac{\omega}{c} \cdot \sqrt{\frac{\omega \cdot \varepsilon_0}{2\varkappa_2}} + \mathrm{j}\,\frac{\omega}{c} \cdot \sqrt{\frac{\omega \cdot \varepsilon_0}{2\varkappa_2}} \tag{1.271}$$

$$\gamma_y^{(2)} = \alpha_y^{(2)} + j\beta_y^{(2)} \approx \sqrt{\frac{\omega \cdot \mu_0 \cdot \varkappa_2}{2}} + j\sqrt{\frac{\omega \cdot \mu_0 \cdot \varkappa_2}{2}} . \tag{1.272}$$

With this we obtain, if medium (1) air and medium (2) are good conductors, a representation of the propagation factors of the field components which is written in a representative form for the $E_z$-component

$$\underline{E}_{z_1} = \underline{A}_1 \cdot e^{-\left(\frac{\omega}{c}\sqrt{\frac{\omega\varepsilon_0}{2\varkappa_2}} \cdot y + \frac{\omega^2 \cdot \varepsilon_0}{2c \cdot \varkappa_2} \cdot z\right)} \cdot e^{j\left(\omega t - \beta \cdot z + \frac{\omega}{c}\sqrt{\frac{\omega \cdot \varepsilon_0}{2\varkappa_2}} \cdot y\right)} \tag{1.273}$$

$$\underline{E}_{z_2} = \underline{A}_2 \cdot e^{\left(\sqrt{\frac{\omega \cdot \mu_0 \cdot \varkappa_2}{2}} \cdot y - \frac{\omega^2 \cdot \varepsilon_0}{2c \cdot \varkappa_2} \cdot z\right)} \cdot e^{j\left(\omega \cdot t - \beta \cdot z + \sqrt{\frac{\omega \cdot \mu_0 \cdot \varkappa_2}{2}} \cdot y\right)} . \tag{1.274}$$

The wave slopes toward the boundary surface and propagates here in the $z$-direction. The field amplitudes decrease toward the outside in the $y$-direction with a finite $\kappa_2$). The wave penetrates the metal somewhat. We recognize in Eq. (1.274) that for the depth at which the field amplitudes die down to the $e$th part of their value at the junction plane, the known depth of penetration yields

$$d = \frac{1}{\alpha_y^{(2)}} = \sqrt{\frac{2}{\omega\mu\varkappa_2}} . \tag{1.275}$$

We obtain the constant amplitude surfaces of the wave by setting the exponent of the e-function equal to a constant with a real argument in Eq. (1.273) or Eq. (1.274). For the upper medium (1) air, we obtain with

$$\frac{\omega}{c}\sqrt{\frac{\omega\varepsilon_0}{2\varkappa_2}}\, y + \frac{\omega^2\varepsilon_0}{2c\varkappa_2}\, z = \text{const.}$$

and

$$\tan\vartheta_1 = -\sqrt{\frac{\omega\varepsilon_0}{2\varkappa_2}} . \tag{1.276}$$

with

$$\frac{\omega}{c}\sqrt{\frac{\omega\varepsilon_0}{2\varkappa_2}}\, y - \frac{\omega}{c} \cdot z = \text{const.}$$

for the inclination angle of this plane with the boundary plane

$$\tan\vartheta_2 = \sqrt{\frac{2\varkappa_2}{\omega\varepsilon_0}} = -\frac{1}{\tan\vartheta_1} . \tag{1.277}$$

The amplitude and phase planes in medium (1) are, therefore, perpendicular to one another. At a very high conductivity $\kappa_2$ the surface wave moves in the medium (1) air almost parallel toward the junction and then penetrates the conductor without reflection almost perpendicularly because it is very sharply attenuated. In free space, the wave is only slightly attenuated, and essentially less so in the $z$-direction than in the $y$-direction. The phase velocity of the surface wavelength of the boundary surface is then practically equal to the speed of light.

In Fig. 1.29a the field relationships at the boundary plane between the medium (1) air and a well conducting medium are outlined. The magnetic field intensity continually crosses over from medium (1) to medium (2). In the conducting medium (2) the electromotive forces induced by the magnetic field intensity with a finite conductivity $\kappa_2$ have currents in the $z$-direction as a result, and, therefore, a horizontal field intensity $E_z$. Because of the stability of this tangential electrical field intensity at the boundary surface a horizontal electrical field intensity is also present in medium (1).

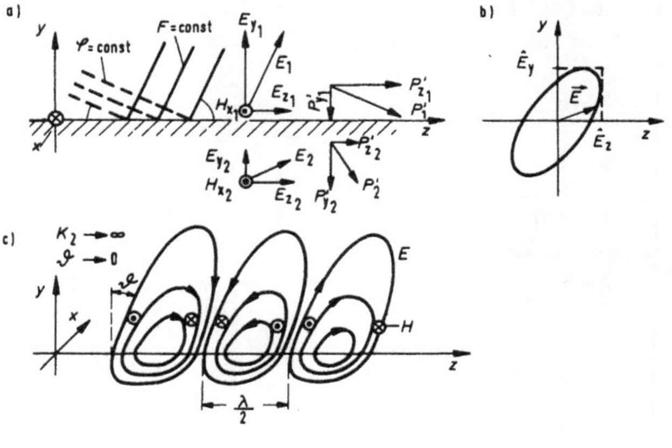


**Fig. 1.29** Plane $E_{oo}$-surface waves (Zenneck-wave)

Following from the stability of the normal part of the current density at the boundary plane, we have

$$j\omega\varepsilon_0 \cdot \underline{E}_{y_1} = \varkappa_2 \cdot \underline{E}_{y_2} \quad \text{or} \quad \frac{\underline{E}_{y_2}}{\underline{E}_{y_1}} = \frac{j\omega\varepsilon_0}{\varkappa_2} \tag{1.278}$$

[compare with Eqs. (1.253) and (1.256)].

For high conductivity $\kappa_2 \gg \omega\varepsilon_0$, then $E_{y_2} \ll E_{y_1}$ and $E_{y_2} \ll E_{z_2}$.

The charge per unit area transported by the wave in the $z$-direction is given by the component $p'_z$ of the Poynting vector; therefore, for medium (1) by

$P'_{z_1} = -E_{y_1} \cdot H_{x_1}$. A small portion of the charge which is drained into the conductor is determined by $P'_{y_2} = E_{z_1} \cdot H_{x_1}$. The resulting Poynting vector $\vec{P}'_1$ in medium (1) is, therefore, essentially directed in the $z$-direction with a small tendency towards medium (2). In the conducting medium (2), the resulting Poynting vector $\vec{P}'_2$ of the equalizing current is almost perpendicular to the boundary surface with a slight tendency towards the $z$-direction. As a result of the energy drain in the conducting medium, the amplitude of the surface wave decreases in the $z$-direction.

For the relation between $E_{z_1}$ and $E_{y_1}$ at the conductor surface, we obtain the following from Eqs. (1.252) and (1.253)

$$\underline{E}_{z_1} \approx \sqrt{\frac{j\omega\varepsilon_0}{\varkappa_2}}\,\underline{E}_{y_1} = \sqrt{\frac{\omega\varepsilon_0}{\varkappa_2}} \cdot e^{j\pi/4} \cdot \underline{E}_{y_1} = (1 + j) \cdot \sqrt{\frac{\omega\varepsilon_0}{2\varkappa_2}} \cdot \underline{E}_{y_1} \,. \tag{1.279}$$

The field components $E_{y_1}$ and $E_{z_1}$, which are perpendicular to one another, have, therefore, a phase difference of 45° with respect to one another. With

$$E_{z_1} = \hat{E}_{z_1} \cdot \sin\omega t \tag{1.280}$$

and

$$E_{y_1} = \hat{E}_{y_1} \cdot \sin\left(\omega t - \frac{\pi}{4}\right) = \frac{\hat{E}_{y_1}}{\sqrt{2}}\,(\sin\omega t - \cos\omega t)\,, \tag{1.281}$$

whereby

$$\hat{E}_{y_1} = \sqrt{\frac{\varkappa_2}{\omega\varepsilon_0}} \cdot \hat{E}_{z_1}\,, \tag{1.282}$$

we obtain

$$E_{y_1} = \frac{\hat{E}_{y_1}}{\sqrt{2}}\left(\frac{E_{z_1}}{\hat{E}_{z_1}} - \sqrt{1 - \left(\frac{E_{z_1}}{\hat{E}_{z_1}}\right)^2}\right). \tag{1.283}$$

From this, we obtain

$$\left(\frac{E_{z_1}}{\hat{E}_{z_1}} - \sqrt{2}\,\frac{E_{y_1}}{\hat{E}_{y_1}}\right)^2 + \left(\frac{E_{z_1}}{\hat{E}_{z_1}}\right)^2 = 1\,. \tag{1.284}$$

According to Eq. (1.284), the vector of the resulting electrical field intensity at the conductor surface describes a sloped ellipse in the $yz$-plane (Fig. 1.29b); thus, the electrical field is elliptically polarized.

In Fig. 1.29c, the field pattern of this Zenneck-wave is outlined. We can see the rotation of the $E$-vector at a point on the conductor surface, following the tangent directions of the $E$-lines that pass by this point at $v_p = c$.

If medium (2) is ideally conductive ($\kappa \to \infty$), then the wave propagates as a plane wave in the non-conducting medium parallel to the conductor surface.

If both media are non-conductive, then the wave comes in at an oblique or sloped angle to medium (1), under the Brewster angle, at which no reflection occurs on the boundary surface, and then penetrates medium (2), whereby the law of reflection is satisfied.

The propagation of a surface wave with cylindrical field symmetry – cylinder axis in the $y$-direction – and a radial direction of propagation is also possible along a plane boundary surface (field components $H_\varphi, E_\rho, E_y$ in cylindrical coordinates). The dependence of the field components upon the radial coordinates $\rho$ may now be described by the Hankel functions, which, however, can already be represented as exponential functions for small values of $\rho$ ($\gamma_\rho > 1$).

In order to characterize surface waves, a *surface impedance* $\underline{Z}_s$ has been defined as the relationship between a tangential component of the electrical field intensity at the boundary surface and the component of the magnetic field intensity which is perpendicular to it.

For the Zenneck-wave discussed above, we obtain

$$\underline{Z}_s = R_s + jX_s = \left[\frac{\underline{E}_{z_1}}{\underline{H}_{x_1}}\right]_{y=0} = -\frac{j\gamma_y^{(1)}}{\omega\varepsilon_0} = (1 + j) \cdot \sqrt{\frac{\omega\mu_0}{2\varkappa_2}} . \tag{1.285}$$

For a conductor with finite conductivity $\kappa$, the surface impedance is inductive ($X_s > 0$). An $E$-surface wave can be directed along such a conductor surface (above example, $E_z \neq 0$). In order to conduct $H$-waves, a capacitative surface impedance is necessary.

For the ideally conductive boundary surface ($\kappa_2 \to \infty$) the surface impedance becomes $Z_s = 0$ with $\gamma_y^1 = 0$, and for the purely dielectric boundary layer, we obtain

$$Z_s = \sqrt{\frac{\mu_0}{\varepsilon_0 + \varepsilon_2}} .$$

The larger the imaginary parts $X_s$ of the surface impedance and frequency are, the greater the voltage drop of the field intensities will be with the distance from the junction, which means the surface wave will be more narrowly attached to the boundary surface. With this, however, the wave phase velocity along the junction decreases. With a larger real part $R_s$ of the surface impedance, the inclination angle which forms the wave fronts of the surface wave with the junction plane, becomes smaller and, therefore, the phase velocity of the wave along the boundary surface becomes higher. With an inductive $X_s$, the phase velocity is lessened, and with a capacitative $X_s$, it is increased. For $R_s \gg X_s, v_p > c$ (rapid waves) and for $R_s \ll X_s, v_p < c$ (slow waves). The attenuation of the wave in the $z$-direction increases with $R_s$ and $X_s$.

We can attain an increase in the surface reactance $X_s$ by thinly stratifying a metallic surface with dielectric material, or by providing it with periodically repeated grooves transverse to the direction of propagation (synthetic dielectric). In the first case, we obtain an inductive surface impedance; the periodically structured metal surface yields areas of capacitative or inductive surface impedances, each according to the relationship between the depth of the groove and the wavelength.

### 1.8.2 Axial Cylindrical Surface Waveguides

#### *1.8.2.1 Metallic Single-Wire Circuit, Sommerfeld Circuit*

In 1899, A. Sommerfeld theoretically demonstrated that electromagnetic waves can propagate along a straight metal wire of finite conductivity. The energy is directed along the wire in the form of surface waves. We distinguish here between $E$-waves and $H$-waves. With $E$-waves only the electrical field shows a component in the direction of the wire axis, and with $H$-waves only the magnetic field does so.

The wave investigated by Sommerfeld is the dynamically balanced $E_{00}$-wave (similar to the Zenneck-wave) with the field components $E_\rho, E_z, H_\varphi$. It is also designated as the line shaft, or the *Sommerfeld single-wire circuit.* In Fig. 1.30 the path of its electrical and magnetic lines of force are outlined. The field amplitudes are strongly attenuated in the radial direction and only slightly attenuated in the axial direction. The wire conductor current is returned to the adjacent free space by the shift currents. With all other $E$- and $H$-wave modes of the single-wire circuit (so-called spurious emission, investigated by D. Hondros), the field runs predominantly in the conductor interior. Therefore, they are also strongly attenuated in the direction of propagation and possess no practical significance.

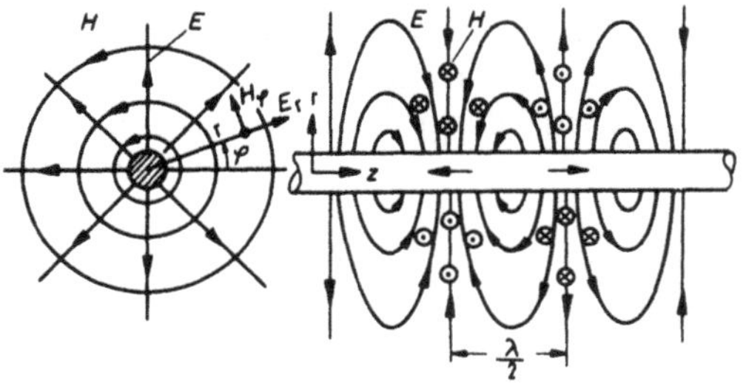


**Fig. 1.30** Cylindrical $E_{oo}$-surface wave (Sommerfeld-wave)

For the propagation constant $\gamma$ of the single-wire circuit, the following is valid:

$$\gamma = \alpha + j\beta = \frac{2\pi j}{\lambda_0}\sqrt{1 + \left(\frac{\lambda_0}{2\pi \underline{r}_0}\right)^2} \approx \frac{2\pi j}{\lambda_0}\left[1 + \frac{1}{2}\left(\frac{\lambda_0}{2\pi \underline{r}_0}\right)^2\right] \quad \text{for} \quad |\underline{r}_0| > \lambda_0 \,. \tag{1.286}$$

Here we designate the complex magnitude $\underline{r}_0$ as the boundary radius, since for $r > r_0$ the field surrounding the wire decreases exponentially and the transferred energy is essentially concentrated in the area $r < r_0$. The value of $\underline{r}_0$ can be obtained from the relation

$$\frac{r_0}{a} = 0{,}279 \left(\frac{\lambda_0}{\sqrt{a \cdot d}}\right)^{1{,}07}, \tag{1.287}$$

$a$ = radius of the wire, $d$ = depth of penetration.

We then obtain the phase angle $\varphi$ of $\underline{r}_0 = r_0 \cdot e^{j\varphi}$ from

$$\varphi = \frac{\pi}{8} + \frac{\pi}{16 \cdot \ln\left(\frac{1{,}125 \cdot r_0}{a} - \frac{1}{2}\right)} \,. \tag{1.288}$$

Therefore, we obtain the attenuation and phase constants, from Eq. (1.286), as

$$\alpha = \frac{\lambda_0}{4\pi r_0^2} \sin 2\varphi \tag{1.289}$$

and

$$\beta = \frac{2\pi}{\lambda} = \frac{2\pi}{\lambda_0}\left[1 + \frac{1}{2}\left(\frac{\lambda_0}{2\pi r_0}\right)^2 \cos 2\varphi\right] \,. \tag{1.290}$$

With higher frequency the attenuation of the Sommerfeld-wave increases and its sidelobe length decreases. The attenuation lessens with a larger wire radius, whereby, obviously, the sidelobe length increases. Its practical application is limited by this since the conductor diameter or the sidelobe length becomes too large and, consequently, external disturbances become unduly high. A favorable compromise transmission line is obtained in the frequency range of approximately 2-6 GHz.

Wave propagation follows in the axial direction with a phase velocity that is virtually equal to the speed of light.

At a greater distance $r$ from the conductor follows the sidelobe dependence of the field magnitudes according to the function $e^{\gamma_r \cdot r}$. For suppressing the sidelobe, it is necessary that $\gamma_r$ possess a real component. This is given for a finite conductivity of the conductor. A stronger field concentration at the conductor can be attained through a worse conductivity; however, this increases the ohmic losses.

### *1.8.2.2 Metallic Single-Wire with Dielectric Layer*

The sidelobe reduction and, therefore, the existence of the surface wave of a wire can be attained by means other than finite conductivity of the wire. The sidelobe reduction implies a real part of the magnitude $\gamma_r$ in the factor $e^{\gamma_r \cdot r}$ of the radial field dependence. In the case of wave propagation in the $z$-direction with the propagation factor $e^{-\gamma_z \cdot z}$, $\gamma_z = \alpha_z + j\beta_z$ follows from the wave equation

$$\gamma_r = \pm\sqrt{\gamma^2 - \gamma_z^2} = \pm\sqrt{-\beta_0^2 - \gamma_z^2} \quad \text{with} \quad \beta_0 = 2\pi/\lambda_0 = \omega/c \tag{1.291}$$

[compare with Eq. (1.239)].

The magnitude $\gamma_r$ can also become real without the presence of attenuation loss $\alpha_z$ in the conductor with $\alpha_z = 0$ and $\gamma_r = \pm\sqrt{(\beta_z^2 - \beta_0^2)}$ for

$$\beta_z > \beta_0, \quad \text{that means} \quad \beta_z = 2\pi/\lambda = \omega/v_p > \omega/c \quad \text{or} \quad v_p < c\,. \tag{1.292}$$

It is, therefore, necessary that the phase velocity $v_p$ of the wave be less than the speed of light $c$. One possible way to do this, namely covering the metallic wire with a thin dielectric layer, was theoretically dealt with in 1907 by F. Harms; its practical application, however, was first investigated in 1950 by G. Goubau. Therefore, we call this circuit the Harms-Goubau circuit. As in the case of the Somerfeld -wire, the dynamically balanced $E_{00}$-wave (Fig. 1.30) is also the only wave form which can yield low attenuation.

Through the dielectric stratification of the metal wire, the boundary radius $r_0$ and, thereby, also the field length are externally reduced. The boundary radius now becomes real. It can be approximately calculated for small layer thicknesses $s \ll \lambda_0/\epsilon_r$ (from (G. Piefke and M. Lohr) according to the relation

$$r_0 \approx \frac{1{,}8}{d_i}\left[\frac{\lambda_0}{\dfrac{\varepsilon_r - 1}{\varepsilon_r}\cdot \ln\left(1 + \dfrac{d_a - d_i}{d_i}\right)}\right]^{1{,}11} \tag{1.293}$$

$d_i = 2a$ = diameter without layer, $d_a$ = diameter with layer.

The field concentration of the conductor increases with a higher frequency, a smaller diameter of the metal wire, and an increasing intensity and inductivity of the layer. It is true that with this the (ohmic and the often smaller dielectric) attenuation increases, such that an appropriate compromise must be made here

For the Harms-Goubau circuit wire wavelength, the following is valid:

$$\lambda = \frac{\lambda_0}{\sqrt{1 + \left(\dfrac{\lambda_0}{2\pi r_0}\right)^2}} \tag{1.294}$$

and for the oscillation resistance

$$Z_L = 60 \cdot \ln \frac{0{,}68\, r_0}{a}\ \Omega\,. \tag{1.295}$$

We obtain the distributed attenuation of the wire resistance:

$$\alpha_R = \frac{1}{\varkappa \cdot 4\pi \cdot a \cdot d \cdot Z_L} \tag{1.296}$$

and for the distributed attenuation due to dielectric losses:

$$\alpha_D = \frac{\tan\delta_\varepsilon \cdot d \cdot Z_0 \cdot \lambda}{2\varepsilon \cdot a \cdot Z_L \cdot \lambda_0^2}\,. \tag{1.297}$$

In Eqs. (1.295) to (1.297), $a$ signifies the wire radius, $d$ the depth of penetration, and $r_0$ the boundary radius according to Eq. (1.293).

The attenuation of the Harms-Goubau circuit is somewhat higher than that of the Sommerfeld-wire in the case of essentially smaller sidelobe length. Because the boundary radii increase with a lower frequency and, therefore, the disturbance effect, wires with a thin stratification ($s \ll a$) are considered for the transmission of dm- and cm-waves. For m-waves, we must select a thicker stratification $s \approx a$.

Inhomogeneities of a surface circuit, such as contortions of the wire circuit lead to emissions.

The *excitation* of waves can occur by way of a metal funnel arranged coaxially with the wire circuit. This serves as a wave-mode transformer between the wave of the feeder line (coaxial transmission-line wave or hollow waveguide wave) and the surface wave to be excited, and as a matching element between the oscillation resistance of the feeder line and the surface waveguide. The field distribution in the funnel's opening should be as similar as possible to that of the surface wave to be excited. According to a suggestion by H. Kaden, we allow the stratification of the Goubau-circuit to begin first in the opening plane of the funnel.

Wire circuits are used as supply lines to antennas.

### *1.8.2.3 Helical Circuit*

Surface waves possess a field component in the direction of propagation or, respectively, a reduced phase velocity. With the Sommerfeld-circuit, the finite conductivity of the wire is responsible for the occurrence of a horizontal electrical field component, and with the Harms-Goubau circuit, the delay action of the wave through the dielectric stratification is responsible. Another possibility for the formation of field components in the horizontal direction, or for the reduction of the phase velocity, consists of coiling the wire. This is used for the helical circuit (see also helical antenna). In transmission line applications, we choose a large inclination angle of the coil from $\alpha = 70°$ to $80°$ ($\tan \alpha = D/2\pi h$), $a$ = wire radius, $h$ = inclination height, in order to keep the ohmic losses small. In this case of strong inclination, we obtain the same transmission values for the helical circuit as for the Harms-Goubau circuit, if the following relation is valid between the models of both circuits.

$$\frac{1}{2} \operatorname{cotan}^2 \alpha = \left(1 - \frac{1}{\varepsilon_r}\right) \cdot \frac{d}{a} . \tag{1.298}$$

For the helical circuit, we obtain the following proximity relationships for the boundary radius $r_0$, the TEM-mode wavelength $\lambda_L$, the TEM-mode oscillation resistance $Z_L$, and the distributed attenuation $\alpha$ [1.10] for $r_0 \gg a$:

$$\frac{2\pi a}{\lambda_0} \operatorname{cotan} \alpha \approx \frac{a}{r_0} \cdot \sqrt{2 \cdot \ln \frac{1{,}125\, r_0}{a}} \tag{1.299}$$

$$\lambda_L = \frac{\lambda_0}{\sqrt{1 + \left(\frac{\lambda_0}{2\pi r_0}\right)^2}} \approx 1 \tag{1.300}$$

$$Z_L \approx 60 \cdot \ln \frac{0{,}68 \cdot r_0}{a}\ \Omega \tag{1.301}$$

$$\alpha \approx \frac{1}{\varkappa \cdot 4\pi \cdot a \cdot d \cdot Z_0} \cdot \frac{2\pi}{\sin^2 a \cdot \ln 0{,}68 \frac{r_0}{a}} \tag{1.302}$$

for $r_0 \ll a$:

$$\frac{2\pi a}{\lambda_0} \operatorname{cotan} \alpha \approx \frac{a}{r_0} \tag{1.303}$$

$$\lambda = \frac{\lambda_0}{\sqrt{1 + \left(\frac{\lambda_0}{2\pi r_0}\right)^2}} \approx \lambda_0 \cdot \sin \alpha \tag{1.304}$$

$$Z_L \approx 60 \frac{r_0}{2a \cdot \sin \alpha}\ \Omega = 60 \frac{\lambda_0}{4\pi a \cos \alpha}\ \Omega \tag{1.305}$$

$$\alpha = \frac{1}{\varkappa \cdot 4\pi \cdot a \cdot d \cdot Z_0} \cdot \frac{(2\pi)^2 \cdot a \cdot \cos \alpha}{\lambda_0 \cdot \sin^2 \alpha} \tag{1.306}$$

with $a$ the wire radius and $d$ the depth of penetration.

Helical circuits can be realized by coiled metal bands on a cylindrical insulator transmitter.

According to Goubau, a dielectric layer can be substituted for by the periodic configuration of a metal surface transverse to the direction of propagation of the surface wave to be transmitted; for example, in the form of grooves. Such a surface structure represents a synthetic dielectric, and renders the necessary reduction of the phase velocity in guiding a surface wave and forming possible horizontally directed field components. A corresponding wire circuit can be manufactured by buffing a wire surface, or by providing it with thread-shaped grooves or cuts. In circuits with a periodic surface structure, turbulent (spatially harmonic) overtones occur in addition to the dynamically balanced $E$-surface wave, in contrast to the smooth wire circuit, whose intensity can only be maintained as low as possible by suitable sizing of the periodic structure. Additionally, periodically structured surface waveguides only transmit within certain frequency bands.

### *1.8.2.4 Dielectric Circuits*

With dielectric waveguides, we use the complete reflection of electromagnetic waves in an optically thinner medium in order to conduct waves within the more optically dense medium.

The propagation of electromagnetic waves along a homogeneous dielectric circular cylinder was theoretically investigated in 1910 by D. Hondros and P. Debye, and experimentally proven in 1915 by H. Zahn and O. Schriever. Infinitely many field modes are possible. Pure $E$- or $H$-modes, for which only the electrical or magnetic field shows a component in the direction of propagation, are only possible with a dynamically balanced field. Here the electrical lines of force form a concentric circle around the dielectric cylinder in the case of $H_{0n}$-modes, whereby the electrical field intensity at the cylinder surface need not disappear. The magnetic lines of force projected upon the cross-section run in the radial direction. In the case of the $E_{0n}$-modes, the electrical field intensity projected upon a cross section possesses only radial directions. The magnetic field is purely transverse and runs in the radial direction. Additional field modes are possible, which show electrical as well as magnetic field components in the direction of propagation and whose fields are not dynamically balanced. These are designated as $HE_{mn}$-modes when their field pattern is similar to that of an $H$-mode, or as $EH_{mn}$-modes when they resemble an $E$-mode.

Of special interest is the $HE_{11}$-wave as a surface wave of the dielectric circuit. It has no finite critical wavelength, whereas all other field modes possess an upper critical wavelength. The $E_{01}$-mode has the largest critical wavelength with $\lambda_c = 2.61 \cdot \sqrt{(\epsilon_r - 1)} - a$. We choose the diameter $D = a$ of the dielectric filament such that only the surface wave is capable of propagating, which means

$$D \leqq \frac{0{,}766 \cdot \lambda_0}{\sqrt{\varepsilon_r - 1}} \,. \tag{1.307}$$

Figure 1.31 shows the pattern of the lines of force of the $HE_{11}$-wave, which is designated as the dipole wave. The field runs similarly to that of an $H_{11}$-wave in the circular waveguide, whereby the limit conditions of metal are not satisfied at the boundary surface dielectric-air.

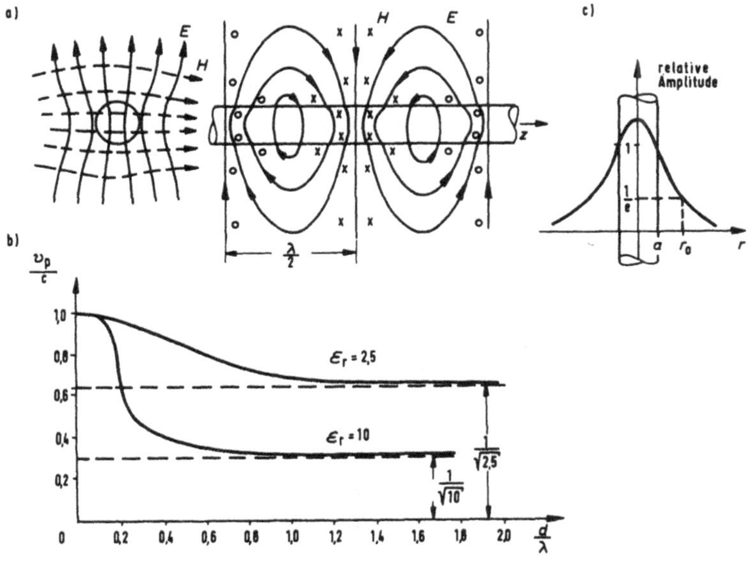


**Fig. 1.31** HE-surface wave (dipole wave) of the dielectric circuit

The attenuation of the dielectric circuit lessens with an ever-decreasing relationship of the diameter to the wavelength $D/\lambda_0$ (reverse of the Sommerfeld-circuit), since an ever-decreasing portion of the energy is transferred to the filament interior. The phase velocity $v_p$ approaches the value $c/\sqrt{(\epsilon_r)}$ for a very large $D/\lambda_0$ and approaches the speed of light $c$ for very small $D/\lambda_0$ (Fig. 1.31c). Therefore the sidelobe length increases with the decreasing relationship $D/\lambda_0$. At $v_p \approx c$ practically all the energy is directed in the exterior, and for $v_p \rightarrow c/\sqrt{(\epsilon_r)}$, in the bar. The sidelobe reduction at the exterior runs according to a Hankel function, which can be approached with increasing distance by an e-function (Fig. 1.31b). We designate the distance from the filament axis as the boundary radius $r_0$, at which the field intensity is lowered to $1/e$ with respect to the value of the filament surface. In the choice of filament diameter, we must deal with a compromise between small attenuation and negligible lobe expansion ($D/\lambda_0 \approx 0.2$-$0.5$). For this reason, the dielectric circuit is not suited for anything larger than cm-waves because the lobe expansion is too large. The strong frequency dependence of the sidelobe expansion is disturbing to a signal transmission for high inductivity of the filament, and therefore of attenuation and phase velocity (dispersion).

As a filament dielectric material, we use, for example, polyethylene with $\epsilon_r = 2.4$ and $\tan \delta = 10^{-4}$.

In order to excite a surface wave on the dielectric circuit, we can allow the circuit to extend into a hollow waveguide connected to a funnel with a circular cross section. With another possibility [1.16], we direct the dielectric filament below 45° through a metal mirror, which is irradiated below 45° by a plane wave formed by a round funnel radiator with a preset lens (see antennas).

In the plane of symmetry of the $HE_{11}$-field distribution (Fig. 1.31a), upon which the electrical lines of force are directed normally and the magnetic lines of force are directed tangentially, a thin and ideally conductive metal plate can be attached without disturbing the field pattern. In this manner, a dielectric circuit may be maintained. We designate such a circuit form as a dielectric *image line* (according to D.D. King). Additional ohmic losses occur in practice through the finite conductivity of the plate. Figure 1.32 shows the cross section arrangement. The plotted monopole antenna with a reflector plate attached at a distance $\lambda/4$ serves the excitation. This circuit is also suited only to the transmission of mm-waves.

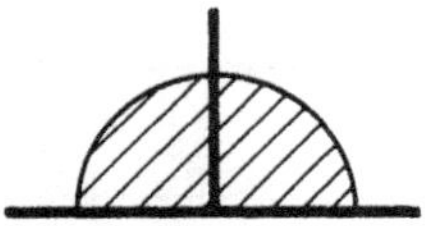

**Fig. 1.32** Dielectric Image Line Cross Section

If, in the case of the dielectrically stratified single-wire circuit, we leave out the metal wire, then we obtain a *dielectric hollow waveguide* (P. Mallach, H.G. Unger, 1955), which also makes a waveguide in the axial direction possible. Now, only displacement currents are present. With these we attain smaller attenuations for the same dimensions at approximately the same field concentration because of the reduced losses in the metal surface. Also, with respect to the dielectric solid wire, we obtain lower attenuation values for a larger field expansion. Since, with solid wire, attenuation and transition-time distortions rise more sharply at high frequencies, the dielectric hollow waveguide is better suited to the transmission of wide frequency bands. Additional dielectric circuits are represented in Fig. 4.13.

# CHAPTER 2

# STRAGGLING PARAMETERS (S-PARAMETERS)

According to network theory, linear antenna networks can be described by parameters, which can be determined by measuring the voltage or the current at the outer terminal pairs with the power off or a short circuit. The usual representations for parameters are the resistance matrix, the conductance matrix, or the iterative matrix.

With high frequency circuits, for which the dimensions of the circuit components are on the same order of magnitude as the wavelength, we run into a few problems. In addition to time, signals are also a function of location; thus, they possess wave character. The line elements exhibit different behavior for various frequencies or lengths and are very often used as circuit components. The input or output of a signal from the antenna network need not consist of two connectors or poles; if the signal is transported by a waveguide (*E*-, *H*-wave), then the connection consists of an aperture (Fig. 2.1). We speak, therefore, more accurately of *N*-gates than of 2 *N*-poles.

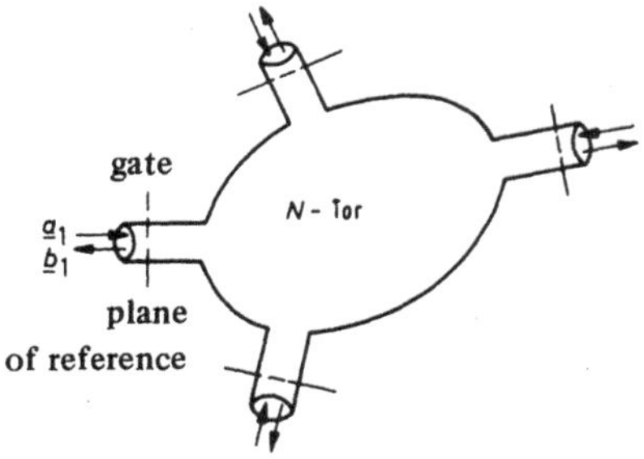


**Fig. 2.1** N-gate

A description of the circuit components can follow with the help of electromagnetic waves, or with equivalent voltage and current waves. With circuits which conduct purely transverse field waves (TEM-circuits), the electrical fields can be clearly assigned equivalent electrical voltage, and the magnetic fields can be assigned current.

In the case of waves with one field component in the direction of propagation (*E*-, *H*-waves), such assignment is also possible, however, not in a clear manner. The transverse-electric and the transverse-magnetic field intensities are connected at a point on the waveguide by a constant, the characteristic wave impedance, and the transfer power is determined by an integral containing only the transverse field components. We therefore introduce equivalent voltage or current waves which are proportional to the total transverse-electrical or transverse-magnetic field intensities.

Many different wave modes can propagate simultaneously on a waveguide. Since, in a lossless circuit, in which more wave modes propagate, the entire transported charge is equal to the sum of all the charges transported by the individual wave modes, a waveguide which propagates *n*-wave modes can be equivalently represented by *n*-TEM-waveguides with corresponding voltage and current waves. The voltage and current are referred to the plane of reference. Every wave mode which transports power independently of all other modes corresponds to a gate (Fig. 2.1).

The description of voltage and current waves, and the derived apparent equivalent resistance, lead to a representation through distinct equivalent-circuit diagrams with concentrated circuit components.

In microwave engineering, rather than with the derived magnitudes of the total voltage and the total current at the gates (and the resistance, conductance, or iterative matrices which result from their help), we work better with the directly measurable amplitude and phase of the incident or reflected waves, and we thereby attain a description of microwave circuits through the *scattering matrix.*

## 2.1 CIRCUIT DESCRIPTION BY STRAGGLING PARAMETERS (S-PARAMETERS)

The behavior of linear antenna networks – or, rather, in the case of correspondingly low modulation for networks considered to be linear – can be described by parameters which are measured at the terminal pairs (gates) of the network. In the case of resistance, conductance, hybrid, or chain parameters, the total voltage (superposition of an incident and reflected voltage wave) and the total current at the gates with the power off or with a short circuit are utilized. At higher frequencies, however, the realization of an exact short circuit, and above all idling power (radiation!), is very difficult because the short circuit usually obtains an inductivity and the idling power is subject to stray inductances. If, with the network to be described, we are dealing with an active element (transistor, active two-port network), then this tends to oscillate for such a connection with short circuit or idling power. Additionally, measuring the total voltage and the total current becomes problematic at higher frequencies.

Therefore, we usually use the so-called *straggling parameters* (S-parameters) for frequencies above approximately 100 MHz, which are measurable in the case of adaptation. The designation straggling parameters originates from the fact that the amplitudes of waves scattered (reflected) on the antenna network can be related to the amplitudes of incident waves. The *scattering matrix* is used to describe active and passive components.

The scattering matrix comprised of the straggling parameters connects the normalized complex amplitudes $\underline{a}$ of the wave moving toward the gates of the network with the normalized complex amplitudes $\underline{b}$ of the reflected wave. Thus, the wave magnitudes $\underline{a}_i$ and $\underline{b}_i$ at gate $i$ of an $N$-gate, $i = 1, 2, \ldots N$, are regulated (power wave according to K. Kurokowa) such that

$$\frac{1}{2}\,\underline{a}_i \cdot \underline{a}_i^* = \text{power available at gate } i \tag{2.1}$$

$$\frac{1}{2}\,\underline{b}_i \cdot \underline{b}_i^* = \text{power leaving gate } i \tag{2.2}$$

Thus, we obtain the relationship between the voltage $U_i$ and the current $I_i$ at gate $i$ [proof Eqs. (2.30) to (2.33)].

$$\underline{a}_i = \frac{\underline{U}_i + \underline{I}_i \cdot \underline{Z}_i^*}{2\sqrt{|\mathrm{Re}(\underline{Z}_i)|}} \tag{2.3}$$

and

$$\underline{b}_i = \frac{\underline{U}_i - \underline{I}_i \cdot \underline{Z}_i^*}{2\sqrt{|\mathrm{Re}(\underline{Z}_i)|}}\,. \tag{2.4}$$

The total voltage $\underline{U}_i$ and current $\underline{I}_i$ are combined for the incident and reflected parts corresponding to

$$\underline{U}_i = \underline{U}_{ih} + \underline{U}_{ir} \tag{2.5}$$

and

$$\underline{I}_i = \underline{I}_{ih} + \underline{I}_{ir}\,. \tag{2.6}$$

We usually assume the random reference oscillation resistance to be real $\underline{Z}_i = Z_0$.

The relationships at a dual-gate (four-pole network) can be seen quite easily (Fig. 2.2). The planes of reference (1) and (2) for the quantities measured at

*Martin Paul, *Schaltungsanalyse mit S-Parameter* (Circuit Analysis with S-Parameters), Heidelberg: Dr. Alfred Huthig Verlag, 1977.

gates 1 and 2 should be located far enough from the interference sources of the dual-gate so that the interference fields are suppressed and a distinct field mode is available. Figure 2.2 shows agreement between the measured arrow directions.

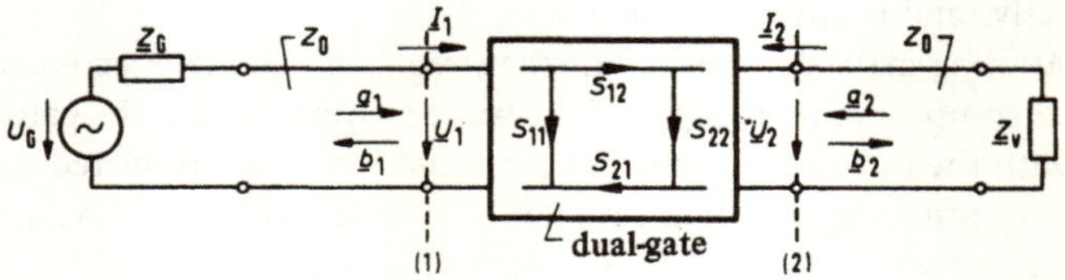


**Fig. 2.2** Dual-gate with agreement of the metered arrow directions

The TEM-mode resistances at gates 1 and 2 should be equal and real. Thus,

$$Z_1 = Z_2 = Z_0 \,. \tag{2.7}$$

Therefore, the following are valid for the dual-gate

$$\underline{a}_1 = \frac{\underline{U}_1 + \underline{I}_1 \cdot Z_0}{2\sqrt{Z_0}} \tag{2.8}$$

$$\underline{b}_1 = \frac{\underline{U}_1 - \underline{I}_1 \cdot Z_0}{2\sqrt{Z_0}} \tag{2.9}$$

$$\underline{a}_2 = \frac{\underline{U}_2 + \underline{I}_2 \cdot Z_0}{2\sqrt{Z_0}} \tag{2.10}$$

$$\underline{b}_2 = \frac{\underline{U}_2 - \underline{I}_2 \cdot Z_0}{2\sqrt{Z_0}} \,. \tag{2.11}$$

Therefore, the linear equations which describe the dual-gate are

$$\underline{b}_1 = \underline{S}_{11}\underline{a}_1 + \underline{S}_{12}\underline{a}_2 \tag{2.12}$$

$$\underline{b}_2 = \underline{S}_{21}\underline{a}_1 + \underline{S}_{22}\underline{a}_2 \tag{2.13}$$

with the magnitudes $\underline{a}_1$ and $\underline{a}_2$ of the incident wave as the independent variable, and $\underline{b}_1$ and $\underline{b}_2$ of the reflected wave as the dependent variable quantity. In matrix notation, the system of equations reads

$$\begin{bmatrix} \underline{b}_1 \\ \underline{b}_2 \end{bmatrix} = \begin{bmatrix} \underline{S}_{11} & \underline{S}_{12} \\ \underline{S}_{21} & \underline{S}_{22} \end{bmatrix} \cdot \begin{bmatrix} \underline{a}_1 \\ \underline{a}_2 \end{bmatrix} \tag{2.14}$$

or, in a general form,

$$[\underline{b}] = [\underline{S}] \cdot [\underline{a}] \,. \tag{2.15}$$

We designate $[S]$ as the S-matrix and its coefficients $\underline{S}_{ik}$ as S-parameters.
Proceeding from Eqs. (2.12) and (2.13) for the dual-gate, we define the S-parameters and their physical characteristics:

By feeding at gate 1 and with $a_2 = 0$; meaning adaptation to gate 2

$$\underline{S}_{11} = \left.\frac{\underline{b}_1}{\underline{a}_1}\right|_{a_2=0} = \text{(inherent) reflectivity at gate 1 with adaptation at gate 2 } (Z_V = Z_0) \tag{2.16}$$

$$\underline{S}_{21} = \left.\frac{\underline{b}_2}{\underline{a}_1}\right|_{a_2=0} = \text{forward transfer or forward transmission factor for transmission from gate 1 to gate 2 with adaptation at gate 2 } (Z_V = Z_0) \tag{2.17}$$

By feeding at gate 2 and with $a_1 = 0$; therefore, adaptation at gate 1, we obtain

$$\underline{S}_{22} = \left.\frac{\underline{b}_2}{\underline{a}_2}\right|_{a_1=0} = \text{(inherent) reflectivity at gate 2 with adaptation at gate 1 } (Z_G = Z_0) \tag{2.18}$$

$$\underline{S}_{12} = \left.\frac{\underline{b}_1}{\underline{a}_2}\right|_{a_1=0} = \text{forward transfer or reverse transmission factor for transmission from gate 2 to gate 1 with adaptation at gate 1 } (Z_G = Z_0) \tag{2.19}$$

The following is generally valid for the S-parameters of an $N$-gate:

$$\underline{S}_{kk} = \left.\frac{\underline{b}_k}{\underline{a}_k}\right|_{a_n = 0 \text{ für } n \neq k} = \text{(inherent) reflectivity of the gate for the case where only gate } i \text{ is fed and all other gates are reflection-free and closed (i.e., all } a_n = 0, \text{ except } n = k\text{), for} \tag{2.20}$$

$$\underline{S}_{ik} = \left.\frac{\underline{b}_i}{\underline{a}_k}\right|_{a_n = 0 \text{ für } n \neq k} = \text{transfer or transmission factor for transfer from gate } k \text{ to gate } i. \tag{2.21}$$

It is advantageous that measurement of S-parameters (e.g., with a directional coupler) follows in the case of a gate terminal adapter.
According to Eq. (2.9), we find by comparing Eqs. (1.94) and (1.95) that $\underline{a}_1 = \underline{U}_{1h}/\sqrt{(Z_0)}$ and, therefore,

$$|\underline{a}_1|^2 = \text{incident power in gate 1 of the network} \tag{2.22}$$

Correspondingly, the following are valid:

$|\underline{b}_1|^2 =$ power reflected at gate 1 (2.23)

$|\underline{a}_2|^2 =$ incident power in gate 2 reflected by the load (2.24)

$|\underline{b}_2|^2 =$ power leaving gate 2 into the load (2.25)

Therefore, the relationships valid for the corresponding terminal adapter are confirmed:

$$|\underline{S}_{11}|^2 = \frac{\text{power reflected at gate 1}}{\text{power entering in gate 1}} = \tag{2.26}$$

$$|\underline{S}_{22}|^2 = \frac{\text{power reflected at gate 2}}{\text{power entering in gate 2}} = \tag{2.27}$$

$$|\underline{S}_{21}|^2 = \frac{\text{power supplied to a load } Z_0 \text{ at gate 2}}{\text{power available from a source with } Z_G = Z_0 \text{ at gate 1}} \tag{2.28}$$

= power transmission ratio at output impedance and ballast resistor $Z_0$.

$$|\underline{S}_{12}|^2 = \frac{\text{power supplied to a load } Z_0 \text{ at gate 1}}{\text{power available from a source with } Z_G = Z_0 \text{ at gate 2}} \tag{2.29}$$

= reverse power transmission ratio at output impedance and ballast resistor $Z_0$.

Still lacking is the proof that the relationships (2.1) and (2.2) between the wavelengths are satisfied, according to Eqs. (2.9) through (2.12).

With $\underline{U}_1 = \underline{U}_G - \underline{I}_1 Z_0$, according to Eq. (2.9)

$$\underline{a}_1 = \frac{1}{2} \cdot \frac{\underline{U}_G}{\sqrt{Z_0}} \tag{2.30}$$

$$\frac{1}{2}\,\underline{a}_1\underline{a}_1^* = \frac{\underline{U}_G^2}{8\,Z_0} \tag{2.31}$$

which was maintained in Eq. (2.1).
With $\underline{U}_2 = -\underline{I}_2 Z_0$, Eq. (2.12) yields

$$\underline{b}_2 = \frac{\underline{U}_2}{\sqrt{Z_0}} \tag{2.32}$$

and

$$\frac{1}{2}\,\underline{b}_2\underline{b}_2^* = \frac{\underline{U}_2}{2Z_0} \tag{2.33}$$

corresponding to Eq. (2.2).
With Eqs. (2.9) and (2.10), we obtain for $\underline{S}_{11}$ (adaptive antenna network with $Z_0$ closed)

$$\underline{S}_{11} = \left.\frac{\underline{b}_1}{\underline{a}_1}\right|_{a_2=0} = \frac{\dfrac{\underline{U}_1}{\underline{I}_1} - Z_0}{\dfrac{\underline{U}_1}{\underline{I}_1} + Z_0} = \frac{\underline{Z}_1 - Z_0}{\underline{Z}_1 + Z_0} = \underline{r}_1 \tag{2.34}$$

With the impedance at gate 1

$$\underline{Z}_1 = \frac{\underline{U}_1}{\underline{I}_1}\,. \tag{2.35}$$

Therefore, we obtain the well known input reflectivity.
Analogously, the following is valid at gate 2:

$$\underline{S}_{22} = \frac{\underline{Z}_2 - Z_0}{\underline{Z}_2 + Z_0} \tag{2.36}$$

with

$$\underline{Z}_2 = \frac{\underline{U}_2}{\underline{I}_2}\,. \tag{2.37}$$

The solution, according to the input impedances $\underline{Z}_1$ at gate 1 and $\underline{Z}_2$ at gate 2, yields

$$\underline{Z}_1 = Z_0\frac{1 + \underline{S}_{11}}{1 - \underline{S}_{11}} \tag{2.38}$$

and

$$\underline{Z}_2 = Z_0 \frac{1 + \underline{S}_{22}}{1 - \underline{S}_{22}} . \qquad (2.39)$$

We obtain the well known transformation [compare with Eq. (1.102)] that formed the basis of the Smith-Chart, and which, therefore, allows us to work comfortably with S-parameters.

For the input reflectivity $\underline{r}'_1$ of the network by random termination with a load impedance $Z_V$ corresponding to a load reflectivity

$$\underline{r}_V = \frac{\underline{Z}_V - Z_0}{\underline{Z}_V + Z_0} \qquad (2.40)$$

we obtain

$$\underline{r}'_1 = \underline{S}_{11} + \frac{\underline{S}_{21} \cdot \underline{S}_{12} \cdot \underline{r}_V}{1 - \underline{S}_{22} \cdot \underline{r}_V} . \qquad (2.41)$$

With reversible networks (containing only passive elements and demonstrating no directivity), the following holds true:

$$\underline{S}_{ik} = \underline{S}_{ki} . \qquad (2.42)$$

Therefore, the S-matrix is symmetrical; the S-matrix $[S]$ is then equal to its transposition (transposed matrix), which follows from the S-matrix by exchanging its rows and columns

$$[\underline{S}] = [\underline{S}]^T . \qquad (2.43)$$

If the relationship is also satisfied for $i = k$, then we are dealing with a *symmetrical network.*

In the case of a *lossless network,* the incoming power is equal to the power displaced according to the theorem of energy conservation; therefore,

$$\Sigma|\underline{a}_n|^2 = \Sigma|\underline{b}_n|^2 . \qquad (2.44)$$

The S-matrix is then *unitary*; hence, the following is valid:

$$[E] - [\underline{S}^*]^T [\underline{S}] = 0 \qquad (2.45)$$

where [E] = the unit matrix, $[\underline{S}^*]$ = transposed conjugated complex S-matrix (hermetic conjugated S-matrix).

With a lossless dual-gate, we obtain from Eq. (2.44)

$$|\underline{S}_{21}|^2 = 1 - |\underline{S}_{11}|^2 \qquad (2.46)$$

$$|\underline{S}_{12}|^2 = 1 - |\underline{S}_{22}|^2 \qquad (2.47)$$

and

$$\underline{S}_{11} \cdot \underline{S}_{12}^* + \underline{S}_{21} \cdot \underline{S}_{22}^* = 0 \qquad (2.48)$$

$$\underline{S}_{11}^* \cdot \underline{S}_{12} + \underline{S}_{21}^* \cdot \underline{S}_{22} = 0\,. \qquad (2.49)$$

With $\underline{S}_{12} = \underline{S}_{21}$, the following is valid:

$$|\underline{S}_{11}| = |\underline{S}_{22}| \qquad (2.50)$$

and

$$\varphi_{11} + \varphi_{22} = 2 \cdot \varphi_{12} \pm 2\pi\,, \qquad (2.51)$$

whereby the phase angles of the S-parameters in question are $\varphi$, corresponding to

$$\underline{S}_{11} = |\underline{S}_{11}| \cdot e^{j\varphi_{11}}, \quad \underline{S}_{22} = |\underline{S}_{22}| \cdot e^{j\varphi_{22}}, \quad \underline{S}_{12} = \underline{S}_{21} = |\underline{S}_{12}| \cdot e^{j\varphi_{12}}\,. \qquad (2.52)$$

If the four-pole network is also symmetrical, then $\varphi_{11} = \varphi_{22}$.

In the case of a *dissipative network,*

$$\Sigma|\underline{b}_n|^2 < \Sigma|\underline{a}_n|^2 \qquad (2.53)$$

and

$$[E] - [\underline{S}^*]^T \cdot [\underline{S}] > 0\,. \qquad (2.54)$$

The difference between the incident and reflected power is the dissipation loss of the network.

As a simple example of application, the S-matrix for a transverse conductance $\underline{Y}$ is calculated (Fig. 2.3).

$\underline{S}_{11}$ can be considered the input reflectivity associated with the input conductance $\underline{Y}_1$ of the dual-gate at a terminal of gate 2 with the reference conductance $Y_0$. The following is valid:

$$\underline{S}_{11} = \left.\frac{\underline{b}_1}{\underline{a}_1}\right|_{a_2=0} = \underline{r}_1 = \frac{Y_0 - \underline{Y}_1}{Y_0 + \underline{Y}_1} = \frac{Y_0 - \underline{Y} - Y_0}{Y_0 + \underline{Y} + Y_0} = -\frac{\underline{Y}}{2\,Y_0 + \underline{Y}} = -\frac{\underline{Y}/Y_0}{\underline{Y}/Y_0 + 2}\,. \qquad (2.55)$$

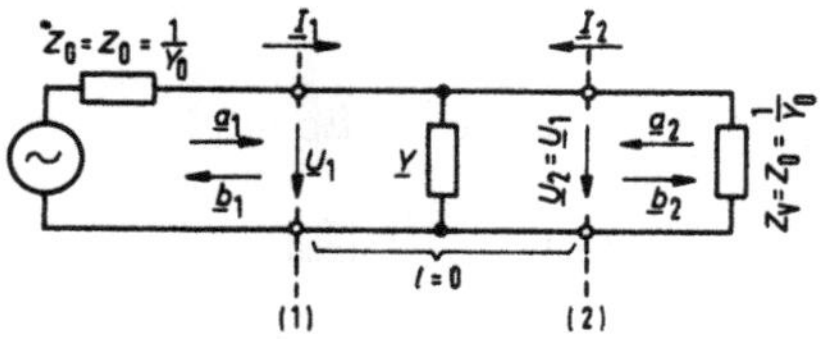


**Fig. 2.3** Calculation of the S-parameters for transverse conductance $\underline{Y}$

Correspondingly, we obtain with a termination of the source at gate 2 and the charge $Y_0$ at gate 1 for $\underline{S}_{22}$

$$\underline{S}_{22} = \left.\frac{\underline{b}_2}{\underline{a}_2}\right|_{a_1 = 0} = \underline{S}_{11} = -\frac{\underline{Y}/Y_0}{\underline{Y}/Y_0 + 2}\,. \tag{2.56}$$

With Eqs. (2.8) to (2.11) this is

$$\underline{U}_1 = \underline{U}_2 = (\underline{a}_1 + \underline{b}_1)\cdot\sqrt{Z_0} = (\underline{a}_2 + \underline{b}_2)\cdot\sqrt{Z_0} \tag{2.57}$$

and, therefore,

$$\underline{a}_1 + \underline{b}_1 = \underline{a}_2 + \underline{b}_2\,. \tag{2.58}$$

Therefore, we obtain for $\underline{S}_{12}$ and $\underline{S}_{21}$

$$\underline{S}_{12} = \left.\frac{\underline{b}_1}{\underline{a}_2}\right|_{a_1 = 0} = \left.\frac{\underline{a}_2 + \underline{b}_2}{\underline{a}_2}\right|_{a_1 = 0} = 1 + \underline{S}_{11} = \frac{2\,Y_0}{2\,Y_0 + \underline{Y}} = \frac{2}{2 + \underline{Y}/Y_0} \tag{2.59}$$

$$\underline{S}_{21} = \left.\frac{\underline{b}_2}{\underline{a}_1}\right|_{a_2 = 0} = \left.\frac{\underline{a}_1 + \underline{b}_1}{\underline{a}_1}\right|_{a_2 = 0} = 1 + \underline{S}_{11} = \underline{S}_{12} = \frac{2}{2 + \underline{Y}/Y_0}\,. \tag{2.60}$$

Therefore, the S-matrix for the parallel conductance $Y$ reads

$$\underline{S} = \frac{1}{2\,Y_0 + \underline{Y}}\begin{pmatrix} -\underline{Y} & 2\,Y_0 \\ 2\,Y_0 & -\underline{Y} \end{pmatrix}. \tag{2.61}$$

## 2.2 ATTENUATION COEFFICIENTS OF DUAL GATES*

The *operative attenuation* for a four-pole network is defined as

$$\frac{a_B}{\mathrm{dB}} = 10 \log \frac{P_{G_V}}{P_V} \tag{2.62}$$

*Klein and Motz, "Mehrtortheorie," (Multiple-Gate Theory) in C. Rint, ed., *Handbuch fur Hochfrequenz- und Elektro-Techniker* (Handbook for High Frequency and Electronic Technicians), vol. 2, p. 193 ff.

with $P_{G_V}$ = available generator power = maximum effective power from the generator of a complex load resistance $\underline{Z}_G^*$ conjugated to its internal resistance $\underline{Z}_G$

$P_V$ = active input power from load resistance $Z_V$.

For the input adapter to the four-pole network (gate 1) with $\underline{Z}_G = \underline{Z}_1$ we obtain

$$\frac{a_B}{\text{dB}} = 20 \log \frac{1}{|\underline{S}_{21}|} + 20 \log |1 - \underline{r}_V \cdot \underline{S}_{22}| + 20 \log \frac{1}{\sqrt{1 - |\underline{r}_V|^2}} . \quad (2.63)$$

For the *insertion loss* of a four-pole network, the following is valid:

$$\frac{a_E}{\text{dB}} = 10 \log \frac{P'_V}{P_V} , \quad (2.64)$$

with $P'_V$ = power released by the generator without the interposition of the four-pole network at the receiver ($\underline{Z}_V$).

$P_V$ = power that reaches the receiver via the four-pole network.

With $Z_G = \underline{Z}_1$, with an adaptive generator, we obtain

$$\frac{a_E}{\text{dB}} = 20 \log \frac{1}{|\underline{S}_{21}|} + 20 \log |1 - \underline{r}_V \underline{S}_{22}| . \quad (2.65)$$

Thus, with an adaptive receiver,

$$\frac{a_B}{\text{dB}} = \frac{a_E}{\text{dB}} = 20 \log \frac{1}{|\underline{S}_{21}|} . \quad (2.66)$$

Also, according to Eq. (2.28), with adaptation at the input and output and $Z_G = Z_V = Z_0$ we obtain the attenuation of the four-pole network (Fig. 2.4).

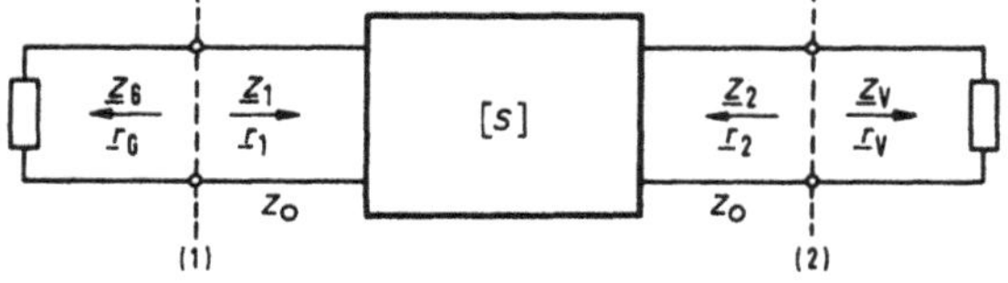


**Fig. 2.4** Defining reflectivity

$$a = \ln\sqrt{\frac{P_1}{P_V}} = \ln\frac{1}{2}\left|\frac{\underline{U}_G}{\underline{U}_V}\right| = \ln\frac{1}{|\underline{S}_{21}|} \tag{2.67}$$

with $P_1$ = active power fed into gate 1.

## 2.3 DUAL-GATE POWER GAIN

In the following we specify some relationships that can be obtained for the various dual-gate amplitudes with the help of S-parameters, and which, for example, are used in the design of transistorized amplifiers. In the equations the following are

$$\underline{r}_G = \frac{\underline{Z}_G - Z_0}{\underline{Z}_G + Z_0} =$$ reflectivity of the generator impedance (mis)matching $\underline{Z}_G$ (2.68)

$$\underline{r}_V = \frac{\underline{Z}_V - Z_0}{\underline{Z}_V + Z_0} =$$ reflectivity of the receiver impedance (mis)matching $\underline{Z}_V$ (2.69)

1. *Power gain* $V_K$ = relationship between the effective active power gain $P_V$ supplied to the load at gate 2 and the active power $P_1$ taken from the dual-gate at gate 1 of the signal's source

$$V_K = \frac{P_V}{P_1} = |\underline{S}_{21}|^2 \cdot \frac{1}{1 - |\underline{r}_1|^2} \cdot \frac{1 - |\underline{r}_V|^2}{|1 - \underline{r}_V \cdot \underline{S}_{22}|^2} \tag{2.70}$$

here $\underline{r}_1$ is the input reflectivity of the dual-gate with

$$\underline{r}_1 = \underline{S}_{11} + \frac{\underline{S}_{12} \cdot \underline{S}_{21} \cdot \underline{r}_V}{1 - \underline{S}_{22} \cdot \underline{r}_V}\,. \tag{2.71}$$

2. *Transducer power gain* $V_Ü$ = relationship between the effective active power supplied at gate 2 to the receiver and the available (with adaptation) power $P_{G_V}$ of the signal's source

$$V_{\ddot{U}} = \frac{P_V}{P_{Gv}} = |\underline{S}_{21}|^2 \cdot \frac{1 - |\underline{r}_G|^2}{|1 - \underline{r}_G \cdot \underline{S}_{11}|^2} \cdot \frac{1 - |\underline{r}_V|^2}{|1 - \underline{r}_V \cdot \underline{r}_2|^2} \tag{2.72}$$

with the output reflectivity of the dual-gate

$$\underline{r}_2 = \underline{S}_{22} + \frac{\underline{S}_{12} \cdot \underline{S}_{21} \cdot \underline{r}_G}{1 - \underline{S}_{11} \cdot \underline{r}_G} \quad . \tag{2.73}$$

3. *Unilaterial power gain* $V_{\text{Uu}}$ = power gain of a non-interactive (unilateral) dual-gate. With $S_{12} = 0$, $S_{21} \neq 0$ and, therefore, with $\underline{r}_1 = \underline{S}_{11}$, we obtain

$$V_{\ddot{U}_u} = |\underline{S}_{21}|^2 \cdot \frac{1 - |\underline{r}_G|^2}{|1 - \underline{r}_G \cdot \underline{S}_{11}|^2} \cdot \frac{1 - |\underline{r}_V|^2}{|1 - \underline{r}_V \cdot \underline{S}_{22}|^2} \cdot \tag{2.74}$$

where we have the following

$$V_0 = |\underline{S}_{21}|^2 = V_{\ddot{U}}|_{r_G = r_V = 0} \tag{2.75}$$

$$V_G = \frac{1 - |\underline{r}_G|^2}{|1 - \underline{r}_G \cdot \underline{S}_{11}|^2} \tag{2.76}$$

which take into consideration the reflection losses through a mismatch between the source and the dual-gate input.

$$V_V = \frac{1 - |\underline{r}_V|^2}{|1 - \underline{r}_V \cdot \underline{S}_{22}|^2} \tag{2.77}$$

which takes into consideration the reflection losses through a mismatch between the receiver and the dual-gate output.

4. *Available power gain* $V_V$ = relationship between the available active power $P_{V_V}$ at gate 2 of the dual-gate and the available power of the source $P_{G_V}$ at gate 1.

$$V_V = \frac{P_{V_v}}{P_{G_v}} = |\underline{S}_{21}|^2 \cdot \frac{1 - |\underline{r}_G|^2}{|1 - \underline{r}_G \cdot \underline{S}_{11}|^2} \cdot \frac{1}{1 - |\underline{r}_2|^2} \cdot \tag{2.78}$$

5. *Maximum power gain* $V_{max}$ = gain matched at the input and output gates, i.e., with $\underline{r}_G = \underline{r}_1^*$, $\underline{r}_V = \underline{r}_2^*$.

   Then

$$V_K = V_{\ddot{U}} = V_V = V_{max} \; . \tag{2.79}$$

For absolutely stable dual-gates (see sec. 2.5), we obtain

$$V_{max} = \left|\frac{\underline{S}_{21}}{\underline{S}_{12}}\right| \cdot (k - \sqrt{k^2 - 1})$$

with (2.80)

$$k = \frac{1 + |\underline{\Delta}|^2 - |\underline{S}_{11}|^2 - |\underline{S}_{22}|^2}{2 \cdot |\underline{S}_{12}| \cdot |\underline{S}_{21}|} \geqq 1 \,.$$

Usually, the statement for the gain follows in dB

$$\frac{V}{\text{dB}} = 10 \log V \,. \quad (2.81)$$

## 2.4 POLE FREQUENCY RESPONSE LOCUS REPRESENTATION FOR UNILATERAL POWER GAIN

According to Eq. (2.74), the following is valid for the unilateral power gain (i.e., with a non-interactive dual-gate)

$$V_{\text{Ü}_u} = V_0 \cdot V_G \cdot V_V$$

with (2.82)

$$V_0 = |\underline{S}_{21}|^2, \quad V_G = \frac{1 - |\underline{r}_G|^2}{|1 - \underline{r}_G \cdot \underline{S}_{11}|^2}, \quad V_V = \frac{1 - |\underline{r}_V|^2}{|1 - \underline{r}_V \cdot \underline{S}_{22}|^2} \,.$$

$V_{U_u}$ reaches its maximum when the dual-gate is operated in conjugated complex adaptation: for $r_G = \underline{S}^*_{11}$ with the generator matched with the dual-gate, $V_G = V_{max} = 1/-|S^*_{11}|^2$ and for $\underline{r}_2 = \underline{S}^*_{22}$ with matching of the receiver, $V_V = V_{max} = 1/1 - |\underline{S}^*_{22}|^2$. The adaptative power at the input and output (of a transistor, for example) can be attained by the corresponding networks.

The adaptation problem can be solved easily with the help of the Smith Chart. In order to determine the requisite terminal impedances (reflectivity) for the dual-gate so as to realize the desired gain, we designate (1) pole frequency response locations in the complex plane of the receiver reflectivity $\underline{r}_V$, where the points of all $\underline{r}_V$ vectors lie, for which $V_V$ = const; (2) the pole frequency response locations and in the $\underline{r}_G$ plane, where the points of all $\underline{r}_G$ vectors lie, for which $V_G$ = const. For these pole frequency response locations, we obtain the circles as shown in Fig. 2.5.

The centers of the circles of constant gain lie on the line of the vector $\underline{S}_{11}^*$ ($\underline{S}_{22}^*$) originating at the center of the Smith Chart for the $V_G$ ($V_V$) circles. For the distances from the zero position of the reflectivity chart to the centers of the circles the following is valid:

$$m_i = \frac{\nu_i \cdot |\underline{S}_{ii}|}{1 - |\underline{S}_{ii}|^2 \cdot (1 - \nu_i)} \tag{2.83}$$

and for the circle radii

$$R_i = \frac{\sqrt{1 - \nu_i} \cdot (1 - |\underline{S}_{ii}|^2)}{1 - |\underline{S}_{ii}| \cdot (1 - \nu_i)} \tag{2.84}$$

with $i = 1$ for the $V_G$ circles and $i = 2$ for the $V_V$ circles. In the equations $\nu_i$ is

$$\nu_i = V_i |1 - \underline{S}_{ii}|^2 = \frac{V_i}{V_{i\,max}} \,. \tag{2.85}$$

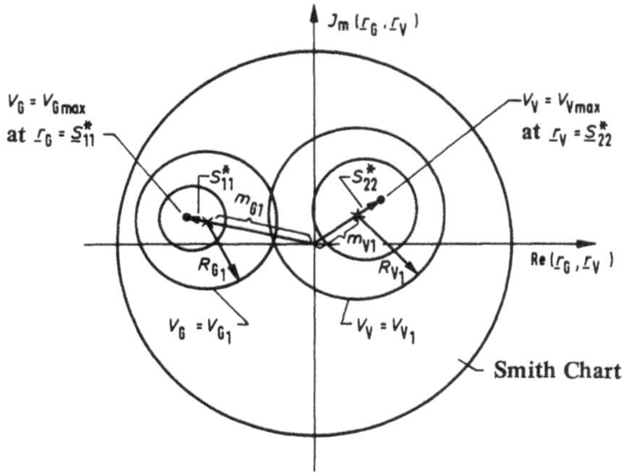


**Fig. 2.5** Circles of constant gain $V_V$, $V_G$ in the reflectivity plane for $r_V$ or $r_G$

## 2.5 STABILITY OF DUAL-GATES

A dual-gate is *stable* if the noise levels caused by a disturbance do not increase and *unstable* if they do increase.

It is *unconditionally (absolutely) stable* at a specified frequency if the real part of the input and output impedances remains positive, which means the portion of the input or output reflectivity, respectively, remains $< 1$ for random passive (i.e., the real part $> 0$) terminations at the input and output. It is *conditionally stable* at a given frequency if the real part of the input and output impedances is positive for some passive source and load impedances; and, therefore, a self-sustained oscillation can occur in certain ranges of the passive terminal resistances.

Oscillations of a dual-gate are only possible if the input impedance at input gate 1 or output gate 2 (or both) possesses a negative real part. However, the dual-gate can operate with stability as well while having a negative real part for these impedances.

For stability, the following condition holds true (compare with Nyquist's stability condition): the pole frequency response location for the sum of the input and source impedances, or of the output or load impedances, must not be allowed to enclose the zero impedance for $f = 0$ to $f \to \infty$.

A dual-gate is, therefore, *unconditionally (absolutely) stable* (at the frequency considered) if

$$|\underline{r}_1| \leqq 1 \text{ for all } |\underline{r}_V| \leqq 1 \text{ and, } \varphi_{r_V} \text{ random} \tag{2.86}$$

and

$$|\underline{r}_2| \leqq 1 \text{ for all } |\underline{r}_G| \leqq 1 \text{ and } \varphi_{r_G} \text{ random} \tag{2.87}$$

therefore, if for all passive load reflectivity $|\underline{r}_V| \leqslant 1$ (receiver) the amount of input reflectivity $|\underline{r}_1| \leqslant 1$, and for all passive load reflectivity $|\underline{r}_G| \leqslant 1$ (generator), then the amount of the input reflectivity $|\underline{r}_2| \leqslant 1$.

In the stability conditions of Eqs. (2.86) and (2.87) (Fig. 2.4), we obtain the following:

$$\underline{r}_1 = \frac{\underline{Z}_1 - Z_0}{\underline{Z}_1 + Z_0} = \text{reflectivity at gate 1} \tag{2.88}$$

$$\underline{r}_2 = \frac{\underline{Z}_2 - Z_0}{\underline{Z}_2 + Z_0} = \text{reflectivity at gate 2} \tag{2.89}$$

$$\underline{r}_G = r_G \cdot e^{j\varphi_{rG}} = \frac{\underline{Z}_G - Z_0}{\underline{Z}_G + Z_0} = \text{generator reflectivity} \tag{2.90}$$

$$\underline{r}_V = r_V \cdot e^{j\varphi_{rV}} = \frac{\underline{Z}_V - Z_0}{\underline{Z}_V + Z_0} = \text{receiver reflectivity} \tag{2.91}$$

$$\underline{Z}_1, \underline{Z}_2 = \text{input impedance at gate 1 or gate 2}$$

Stability testing may begin in the complex planes of the reflectivities $\underline{r}_1$ and $\underline{r}_2$ or in the reflectivity plane for $\underline{r}_G$ and $\underline{r}_V$.

### 2.5.1 Stability Testing in the Reflectivity Planes for $r_1$ and $r_2$

For the input reflectivity of the dual-gate at gates 1 and 2, respectively, to be tested according to the stability conditions from Eqs. (2.86) and (2.87), the following is valid:

$$\underline{r}_1 = \left.\frac{\underline{b}_1}{\underline{a}_1}\right|_{a_2 = 0} = \underline{S}_{11} + \frac{\underline{S}_{12} \cdot \underline{S}_{21} \cdot \underline{r}_V}{1 - \underline{S}_{22} \cdot \underline{r}_V} \tag{2.92}$$

and

$$\underline{r}_2 = \left.\frac{\underline{b}_2}{\underline{a}_2}\right|_{a_1 = 0} = \underline{S}_{22} + \frac{\underline{S}_{12} \cdot \underline{S}_{21} \cdot \underline{r}_G}{1 - \underline{S}_{11} \cdot \underline{r}_G} . \tag{2.93}$$

$\underline{r}_1$ and $\underline{r}_2$ are a function of the S-parameters of the dual-gate; these magnitudes are frequency-dependent. If the stability conditions from Eqs. (2.86) and (2.87) are satisfied for all frequencies, then the dual-gate is absolutely stable over the entire frequency range.

In order the investigate stability, we determine the pole frequency response location in the reflectivity plane of $\underline{r}_1$ for the vector $\underline{r}_1$, where $|\underline{r}_V| = 1$ and $\varphi_{rV}$ is variable; and, in the $\underline{r}_2$ reflectivity plane, the pole frequency response location of the vector for which $|\underline{r}_G| = 1$ and $\varphi_{rG}$ is variable.

With

$$\underline{r}_V = \frac{\underline{S}_{11} - \underline{r}_1}{\Delta - \underline{r}_1 \cdot \underline{S}_{22}} = 1 \tag{2.94}$$

we obtain a circle in the $\underline{r}_1$ plane with its center given by the vector

$$\underline{m}_1 = \underline{S}_{11} + \frac{\underline{S}_{12} \cdot \underline{S}_{21} \cdot \underline{S}_{22}^*}{1 - |\underline{S}_{22}|^2} \tag{2.95}$$

and with a radius of

$$R_1 = \frac{|\underline{S}_{12}| \cdot |\underline{S}_{21}|}{1 - |\underline{S}_{22}|^2} \tag{2.96}$$

This circle is designated as the *stability circle* $|\underline{r}_V| = 1$. All receiver reflectivities with a value $|\underline{r}_V| < 1$ are formed inside this stability circle and all reflectivities $|\underline{r}_V| > 1$ are formed outside of it (since for a dual-gate with $|\underline{S}_{22}| < 1$, $|\underline{r}_V| < 1$).

In Eq. (2.94), Δ is the determinant

$$\Delta = \underline{S}_{11} \cdot \underline{S}_{22} - \underline{S}_{12} \cdot \underline{S}_{21} \, . \tag{2.97}$$

Two examples of such stability circles are represented in Fig. 2.6. Circle A shows a stability circle $|\underline{r}_V| = 1$ for a conditionally stable dual-gate For all passive load reflectivities $|\underline{r}_V| \leqslant 1$ which are depicted inside the stability circle $|\underline{r}_V| = 1$ (where $|\underline{r}_V| < 1$) but outside the shaded region (where $|\underline{r}_1| < 1$), $|\underline{r}_1| > 1$ and the dual-gate is stable, assuming that $|\underline{r}_G| \leqslant 1$ and $|\underline{r}_2| \leqslant 1$ for all passive load reflectivities. In the shaded region the stability condition, Eq. (2.86), is violated with $|\underline{r}_1| > 1$. Circle B is valid for an absolutely stable dual-gate: if the stability circle $|\underline{r}_V| = 1$ lies entirely within circle $|\underline{r}_1| = 1$, then the dual-gate remains stable (at the frequency examined) for all load reflectivities with a value $|\underline{r}_V| \leqslant 1$, if for all $|\underline{r}_G| \leqslant 1$ and $|\underline{r}_2| \leqslant 1$.

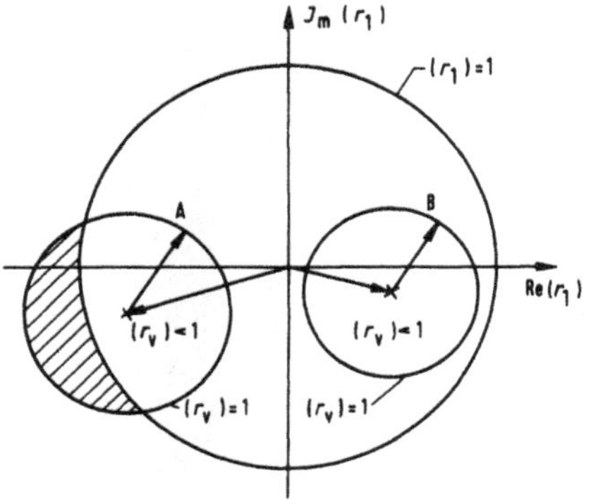


**Fig. 2.6** Stability circle (A, B) in the reflectivity plane for $r_1$ (or $r_2$); A: conditional stability, B: absolute stability

From the absolute stability condition $|m_i| + R_1 \leqslant 1$ for all passive loads with $|\underline{r}_G| \leqslant 1$, we can deduce that

$$|\underline{S}_{12} \cdot \underline{S}_{21}| \leqq 1 - |\underline{S}_{22}|^2 \tag{2.98}$$

and the *stability factor* $k$ must be

$$k = \frac{1 + |\Delta|^2 - |\underline{S}_{11}|^2 - |\underline{S}_{22}|^2}{2 \cdot |\underline{S}_{12} \cdot \underline{S}_{21}|} \geqq 1 \tag{2.99}$$

In the reflectivity plane for $\underline{r}_2$, we obtain correspondingly as a polar frequency response location for the input reflectivities $\underline{r}_2$ at load reflectivities with $|\underline{r}_G| = 1$ and a randon phase angle *a*, *stability circle* $|\underline{r}_G| = 1$ with the central vector

$$\underline{m}_2 = \underline{S}_{22} + \frac{\underline{S}_{12} \cdot \underline{S}_{21} \cdot \underline{S}_{11}^*}{1 - |\underline{S}_{11}|^2} \tag{2.100}$$

and the radius

$$R_2 = \frac{|\underline{S}_{12} \cdot \underline{S}_{21}|}{1 - |\underline{S}_{11}|^2} \; . \tag{2.101}$$

All generator load reflectivities $|\underline{r}_G| < 1$ are depicted inside of the stability circle and all values $\underline{r}_G > 1$ on the outside (since for $|\underline{S}_{11}| < 1$ $|\underline{r}_G| < 1$).

From the stability condition $|\underline{m}_2| + R_2 \leqslant 1$ for all passive loads with $|\underline{r}_G| \leqslant 1$, we now obtain analogously as stability conditions

$$|\underline{S}_{11} \cdot \underline{S}_{21}| \leqq 1 - |\underline{S}_{11}|^2 \tag{2.102}$$

and

$$k = \frac{1 + |\Delta|^2 - |\underline{S}_{11}|^2 - |\underline{S}_{22}|^2}{2 \cdot |\underline{S}_{12} \cdot \underline{S}_{21}|} \geqq 1 \; . \tag{2.103}$$

With this, therefore, a dual-gate is *absolutely stable* if for all frequencies

$$|\underline{S}_{11} \cdot \underline{S}_{21}| \leqq 1 - |\underline{S}_{11}|^2, \; |\underline{S}_{12} \cdot \underline{S}_{21}| \leqq 1 - |\underline{S}_{22}|^2$$

and

$$k = \frac{1 + |\Delta|^2 - |\underline{S}_{11}|^2 - |\underline{S}_{22}|^2}{2 \cdot |\underline{S}_{12} \cdot \underline{S}_{21}|} \geqq 1 \tag{2.104}$$

with the determinant $\Delta = \underline{S}_{11} \cdot \underline{S}_{22} - \underline{S}_{12} \cdot \underline{S}_{21}$.

### 2.5.2 Stability Testing for the Reflectivity Planes of $\underline{r}_G$ and $\underline{r}_V$

The investigation of stability can also begin in the reflectivity plane for $\underline{r}_G$ and $\underline{r}_V$. In order to do so, we determine the pole frequency response locations for the vector $\underline{r}_G$ in the $r_G$ reflectivity plane for which $|\underline{r}_2| = 1$, where $\varphi_{r_2}$ is a variable, and the pole frequency response location in the $\underline{r}_V$ reflection factor plane for $\underline{r}_V$ with $|\underline{r}_1| = 1$, where $\varphi_{r_1}$ is a variable.

We obtain the stability limit for $\underline{r}_V$ by setting $|\underline{r}_1| = 1$ according to Eq. (2.92). The resulting rational fractional function depicts the circle of unit radius $|\underline{r}_1| = 1$ of the $\underline{r}_1$ plane on a circle in the $\underline{r}_V$ plane. As a pole frequency response location of the reflectivity $\underline{r}_V$, we obtain a *stability circle* $|\underline{r}_1| = 1$ in the $\underline{r}_V$ plane with the central vector

$$\underline{m}_V = \frac{\underline{S}_{22}^* - \underline{S}_{11} \cdot \underline{\Delta}^*}{|\underline{S}_{22}|^2 - |\underline{\Delta}|^2} \tag{2.105}$$

and the radius

$$R_V = \frac{|\underline{S}_{12}| \cdot |\underline{S}_{21}|}{|\underline{S}_{22}|^2 - |\underline{\Delta}|^2} \tag{2.106}$$

The stability area $|\underline{r}_1| < 1$, specified inside the circle of unit radius $|\underline{r}_1| = 1$ is depicted outside of the stability circle in the $\underline{r}_V$ plane and all the input reflectivities with $|\underline{r}_1| > 1$ inside the circle, if this does not include the point $\underline{r}_V = 0$ (since $|\underline{r}_1| < 1$ for $|\underline{S}_{11}| < 1$). If this is not the case, the area $|\underline{r}_1| < 0$ is transformed within the interior of the $\underline{r}_V$ circle.

An amplifier operates with stability for every generator impedance with a positive real component if, with a passive termination $|\underline{r}_V| < 1$ of input reflectivity $|\underline{r}_1| < 1$. Therefore, if the stability circle is entirely outside of the circle $|\underline{r}_1| = 1$ and the inside of the stability circle is an unstable area ($|\underline{S}_{11}|$, $|\underline{S}_{22}| < 1$), then the stability condition for the output $|\underline{r}_2| < 1$ does not need to be checked in this case.

Figure 2.7 shows stability circles $|\underline{r}_1| = 1$ for a dual-gate with $|\underline{S}_{11}| < 1$; A only for a conditionally stable dual-gate and B for an absolutely stable dual-gate.

For A the following is valid: for all load reflectivities with a value $|\underline{r}_V| \leqslant 1$ which lie outside the circle $|\underline{r}_V| = 1$ (where $|\underline{r}_V| < 1$) but outside the shaded region (where $|\underline{r}_1| < 1$), then also $|\underline{r}_2| \leqslant 1$. If the stability circle $|\underline{r}_1| = 1$ lies completely outside the circle $|\underline{r}_V| = 1$ (case B), then the dual-gate is stable for all load reflectivities $|\underline{r}_V| \leqslant 1$, at the frequency considered, if for all $|\underline{r}_G| \leqslant 1$ and $|\underline{r}_2| \leqslant 1$.

Correspondingly, when we proceed from Eq. (2.93), we obtain *stability circles* $|\underline{r}_2| = 1$ in the $\underline{r}_G$ reflectivity planes as a pole frequency response location for $|\underline{r}_2| = 1$, where $\varphi_{r_2}$ is a variable.

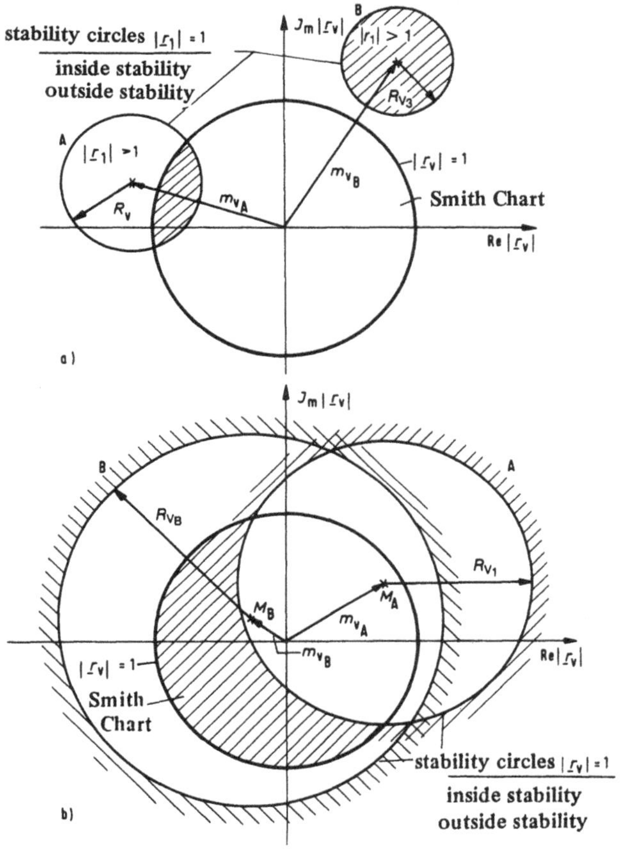


**Fig. 2.7** Stability circles (A,B) in the reflectivity plane for $\underline{r}_V$ (or $\underline{r}_G$); A: conditional stability; B: absolute stability

In this case, the following is valid for the central vector:

$$\underline{m}_G = \frac{\underline{S}_{11}^* - \underline{S}_{22} \cdot \Delta^*}{|\underline{S}_{11}|^2 - |\Delta|^2} \tag{2.107}$$

and for the circle radius:

$$R_G = \frac{|\underline{S}_{12}| \cdot |\underline{S}_{21}|}{|\underline{S}_{11}|^2 - |\Delta|^2} \,. \tag{2.108}$$

In the case of absolutely stable dual-gates all input reflectivities with $|\underline{r}_2| < 1$ are depicted outside the stability circle $|\underline{r}_2| = 1$; all input reflectivities with $|\underline{r}_2| > 1$ are transformed within the interior (since $|\underline{r}_2| < 1$ for $\underline{S}_{22} < 1$).

If the stability conditions are satisfied at all frequencies for all passive load reflectivities $\underline{r}_V$ and $\underline{r}_G$, then the dual-gate is absolutely stable.

The stability circle in the $\underline{r}_V$ plane identifies (in its interior) the values of the load impedances at gate 2 (receiver impedances) which produce a negative real part of the input impedance at gate 1; and, therefore, can cause oscillations. The stability circle in the $\underline{r}_G$ plane specifies (in its interior) the values of the load impedances at gate 1 (source impedances) which yield a negative real part of the input impedance and can cause oscillations.

From the stability conditions with passive load reflectivity $|\underline{r}_G| \leqslant 1$ in the form $||\underline{m}_V| - |R_V|| \geqslant 1$, $||\underline{m}_G| - |R_G|| \geqslant 1$, we also obtain the conditions for absolute stability according to Eq. (2.104).

# CHAPTER 3

# CIRCUIT COMPONENTS IN COAXIAL TRANSMISSION LINE AND HOLLOW WAVEGUIDE TECHNOLOGY

High-frequency lines are not only used for power and signal transmission, but to a great extent for manufacturing circuit components as well. In this chapter, coaxial transmission line components and, above all, hollow waveguide components are described. Stripline components will be dealt with in Chapter 4 and circuit components with gyromagnetic materials (ferrite components) are covered in Chapter 5.

## 3.1 CIRCUIT AND CAVITY RESONATORS

In sec. 1.5.2 we established that for determined values of $l/\lambda$ ($l$ = length of circuit, $\lambda$ = line wavelength) short-circuit or idling circuit components exhibit parallel or series resonances. In the case of resonance, the sum of electrical and magnetic energy is constant over time. A continuous transformation between the electrical and magnetic field energy takes place. We call a resonator single-circuit if at given points in time one type of energy is zero while the other type of energy reaches its maximum value (compare with Fig. 3.6). As the $Q$-factor of the resonator, we then define

$$Q = \frac{\omega_r \cdot W_{max}}{P} = \frac{2\pi \cdot W_{max}}{P \cdot T} \tag{3.1}$$

with

$$\omega_r = 2\pi f_r = \frac{2\pi}{T} = \text{resonant frequency}$$
$$W_{max} = \text{energy received by the resonator}$$
$$P = \text{power loss converted by the resonator}$$

### 3.1.1 Circuit Resonators, Cavity Resonators

Circuit resonators are composed of circuits upon which $L$-waves are converted into standing waves as a result of faulty terminations on both ends (short circuit, idling, or termination with reactance). Figure 3.1 gives a diagrammatic representation, and in Fig. 3.2 a one-sided open resonator (resonances for $l = (2n + 1) \cdot \lambda_r/4$ with $n = 0, 1, 2, \ldots$) and a two-sided short-circuited resonator (resonances for $l = (n + 1) \cdot \lambda_r/2$ with $n = 0, 1, 2, \ldots$) are outlined as examples in coaxial engineering. In order to realize idling power, the dimensions of the circuit's opening must be very small with respect to line wavelength, otherwise radiation occurs.

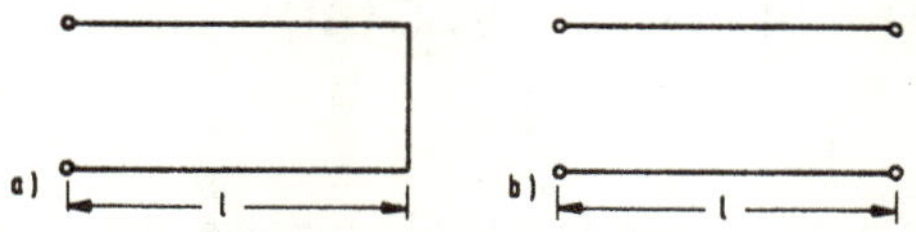


**Fig. 3.1** TEM-circuit resonators: (a) series-resonant circuit at $1 \approx n \cdot \lambda/2$ and parallel-resonant circuit at $1 \approx (2n + 1) \cdot \lambda/2$; (b) parallel-resonant circuit at $1 \approx n\lambda/2$ and series-resonant circuit at $1 \approx (2n + 1) \cdot \lambda/4$

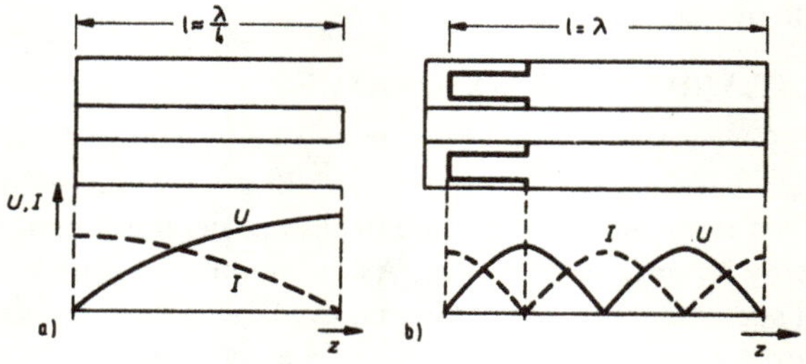


**Fig. 3.2** Coaxial transmission line resonators: (a) $\lambda/4$-resonator, (b) resonator with shorting plunger

The resonant frequency can be changed by altering the length of the circuit. To this end, Fig. 3.2b outlines how a shorting plunger (length $\lambda/4$) with current-free mechanical contacts can be realized.

If the resonator is capacitatively loaded and terminated on all sides by conducting walls, then it is also called a *cavity resonator.* In Fig. 3.3, type models of a coaxial cavity resonator are outlined. The cavity length can be maintained small at the same diameter if at the end of the inner conductor a capacitatively working plate (in Figs. 3.3b, c) is also mounted.

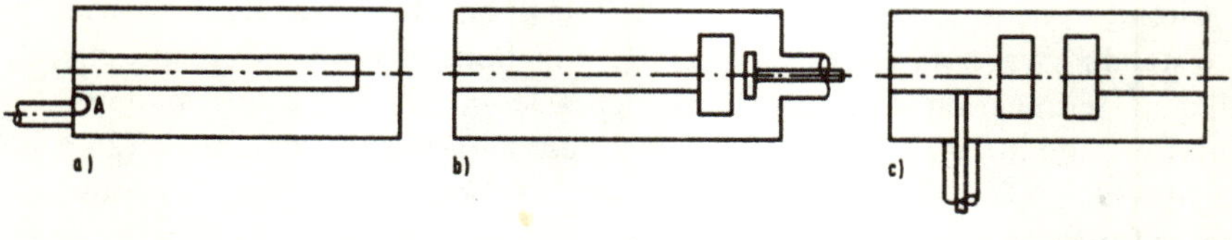


**Fig. 3.3** Coaxial cavity resonators in the vertical cross section with various couplings: (a) inductive, (b) capacitative, (c) galvanic

For a resonator to operate an internally consistent field must be coupled with the field of the waveguide (or waveguides) to be connected. The *resonator coupling* can take place:

(a) from the electrical coupling (predominantly) through the electrical field;
(b) from the magnetic coupling (predominantly) through the magnetic field; or
(c) from the mixed coupling through the electrical and magnetic fields.

The field produced by the coupling element in the resonator must show components in the direction of the desired internally consistent resonator field in the vicinity of the transfer substation.

If the coupling element dimensions are not small compared to wavelength, then we speak of line coupling, which is a mixed coupling. Coupling elements often used are the armature, which predominantly couples magnetically, and the pin, which mainly couples electrically. In waveguide technology, hole and slot coupling are also often used.

In Fig. 3.3 (a, b, and c) various possibilities for coupling a coaxial line to a resonator are represented. Figure 3.3a shows an inductive coupling for which the surface $A$ of a coupling loop is crossed by the magnetic lines of force. For strong coupling, the loop should be placed at a location of relatively high magnetic field intensity or, respectively, large currents, and ought to be directed perpendicular to the magnetic lines of force. In the case of capacitative coupling in Fig. 3.3b, the coupling takes place across the electrical field. We obtain high coupling if the coupling pin is placed at a location of relatively high electrical field intensity, or voltage, and as parallel as possible to the electrical lines of force. In Fig. 3.3c, galvanic coupling is represented. The coupling factor can be changed through the choice of the connection point with the inner conductor. In the case of line coupling, the extended inner conductor of the coaxial feeder cable in the resonator is directed parallel to its neutral wire, and then connected to the resonator wall. Here we obtain a mixed coupling through the electrical and magnetic fields.

The circuit resonator $Q$-factor [Eq. (3.1)] can be approximately calculated as

$$Q = \frac{\omega \cdot \sqrt{L' \cdot C'}}{2\alpha} = \frac{\beta}{2\alpha} = \frac{\pi}{\alpha\lambda} \tag{3.2}$$

with an attenuation constant $\alpha \approx \sqrt{f}$ contingent upon the skin effect, and with $\lambda \approx 1/f$, then $Q \approx \sqrt{f}$.

At higher frequencies, $E$- or $H$-field modes can also occur in circuit resonators.

### 3.1.2 Cavity Resonators (Rectangular and Cylindrical)

If we seal a hollow waveguide with a conducting wall perpendicular to the direction of propagation, then incident and reflected waves are superimposed and create standing waves. The tangential electrical and the normal magnetic field intensities disappear at this wall and at distances of $\lambda_h/2$. In such a nodal plane, a second metal wall can be added without disturbing the field, and thereby we obtain a cavity resonator.

In the case of an excitation of the resonator (for example, through a coupling hole), the field intensity building up within it becomes maximal when the length of the container $c$ (in the $z$-direction) $\lambda_h/2$, or an integral multiple thereof, amounts to

$$c = p \cdot \frac{\lambda_h}{2}, \quad p = 1, 2, 3, \ldots . \tag{3.3}$$

If we insert in Eq. (3.3) the relation for the waveguide wavelength

$$\lambda_h = \lambda_0 / \sqrt{1 - \left(\frac{\lambda_0}{\lambda_g}\right)^2}$$

and solve the equation according to $\lambda_0 = \lambda_r$, then we obtain for the *resonant wavelength* $\lambda_r$

$$\lambda_r = \frac{c}{f_r} = \frac{1}{\sqrt{\left(\frac{p}{2c}\right)^2 + \left(\frac{1}{\lambda_c}\right)^2}} . \tag{3.4}$$

This relation is valid for a resonator with a rectangular cross section as well as for one with a circular cross section, whereby in each case the corresponding equation is to be used for determining the critical wavelength [for the rectangular hollow waveguide Eq. (1.165); for the circular hollow waveguide for $E$-waves Eq. (1.197), and for $H$-waves Eq. (1.205) or Table 1.1]. We must bear in mind that because of the different field modes ($E_{mm}$ and $H_{mn}$) a plurality of resonant frequencies can occur.

In order to maintain oscillation in the empty cavity resonator, we often only need to introduce a small amount of active power, which is dissipated by the wall current losses. The $Q$-factor of a cavity resonator can be calculated by integration over the volume $V$ and the surface $A$ of the cavity

$$Q = \frac{2 \int_{(V)} \underline{\vec{H}} \cdot \underline{\vec{H}}^* \cdot \mathrm{d}V}{d \cdot \int_{(A)} \underline{\vec{H}} \cdot \underline{\vec{H}}^* \cdot \mathrm{d}A} \tag{3.5}$$

$d$ = penetration depth

The energy recorded in the resonator is proportional to its volume and the power loss is proportional to the surface of the inner partition. The $Q$-factor of a cavity resonator is, therefore, approximately proportional to the relationship between volume and surface. In order to achieve a narrowband resonator, we can try to increase the volume by choosing more half-cycles in the $z$-direction. In this instance however, it becomes much more difficult to work with only one resonant frequency. We can suppress disturbing secondary resonances with a favorable excitation of the desired resonance mode, and conducting resistance vanes at a suitable location on the resonator are used to suppress incidental resonances.

Without taking into consideration the losses in both of the reflecting walls, we obtain for the $Q$-factor of the cavity resonator

$$Q_0 = \frac{\pi}{\alpha\lambda} = \frac{v_p}{v_g} \tag{3.6}$$

with $v_p$ = phase velocity

$v_g$ = group velocity

and taking these losses into consideration

$$\frac{1}{Q} = \frac{1}{Q_0} + \frac{4d}{\lambda}\,. \tag{3.7}$$

With cavity resonators, we achieve a very high unloaded $Q$, with well conducting and smooth inner surfaces up to many times $10^4$. The $Q$-factor of cavity resonators decreases proportionally with $1/\sqrt{f}$.

The introduction of dielectric in the resonator increases its resonant wavelength. With complete filling of the dielectric, the following is valid:

$$\lambda_\varepsilon = \lambda \cdot \sqrt{\varepsilon_r}\,. \tag{3.8}$$

By using dielectric material with a high $\epsilon_r$, we can correspondingly reduce the large resonator dimensions necessary for longer wavelengths.

In order to *excite* cavity resonators, the couplings described in sec. 3.1.1 can be used: capacitative coupling with a pin as parallel as possible to the electrical lines of force (corresponding to Fig. 3.4a); inductive coupling with a coupling loop crossed by the magnetic lines of force (Fig. 3.4b); and coupling over the electrical and magnetic fields with strapping at the line coupling. Hole and slot coupling are also customary, especially at higher frequencies. In the case of hole coupling, the coupling follows from a circular hole between the resonator and the connected hollow waveguide. Figure 3.4d shows the path of the lines of force for electrical coupling and magnetic coupling through a hole in the metal wall. Electrical coupling (equivalent representation by an electrical dipole) takes place, for example, through a common hole in the broad sides of the waveguide.

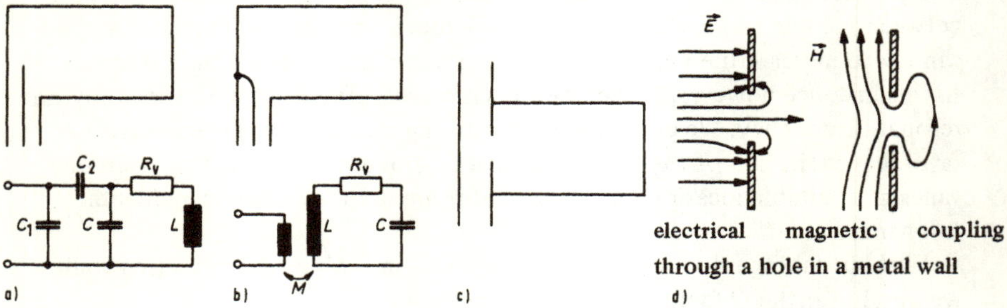


**Fig. 3.4** Cavity resonators: (a) capacitative coupling with probe and simple equivalent circuit; (b) inductive coupling with loops and simple equivalent circuit; (c) coupling with coupling hole; (d) field distribution for a hole coupling

A magnetic coupling (represented by a magnetic dipole) is present with coupling over a hole in a wall of the waveguide's cross-section area. The electrical lines of force of the areas to be coupled should be equally directed at the coupling point. Slot coupling takes place through a common, slot-shaped opening in the resonator and the hollow waveguide. The magnetic lines of force of the hollow waveguide and the resonator should run as parallel as possible to each other at the coupling point and parallel to the slot. Then, the line currents are interrupted through the slot.

The behavior of the resonator is influenced by the coupled load. We will examine the case of a resonator with a matched connecting cable (oscillation resistance $Z_L$) (Fig. 3.5). For the resonator, an equivalent circuit of concentrated structural components can be represented as a parallel-resonant circuit ($L_p$, $C_p$, $R_p$) or as a series-resonant circuit ($L_r$, $C_r$, $R_r$) for a specified reference plane.

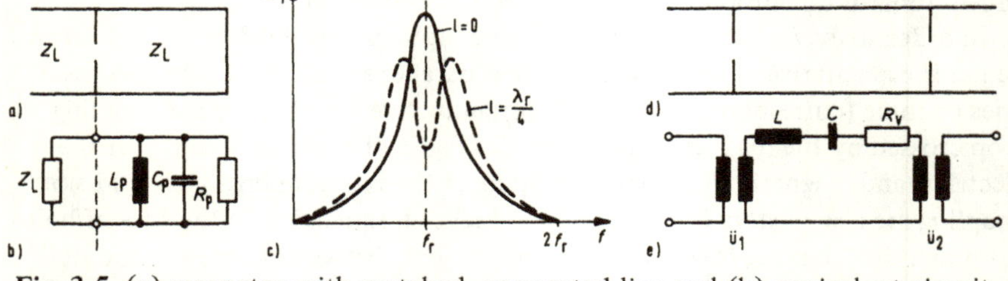


**Fig. 3.5** (a) resonator with matched, connected line and (b) equivalent circuit, (c) input impedance of the loaded resonator ($k > 1$) for $1 = 0$ and $\lambda_v/4$, (d) transit resonator

For the connection between the $Q$-factor of the loaded circuit and unloaded $Q$ ($Q_0$) of the resonator the following is therefore valid:

$$Q = \frac{\omega_r}{\Delta\omega} = \frac{Q_0}{1 + k} \tag{3.9}$$

with

$f_r = \dfrac{\omega_r}{2\pi}$ = resonance frequency with maximum resonator amplitude $A_{max}$,

$\Delta f = \dfrac{\Delta\omega}{2\pi}$ = frequency bandwidth between points of the resonant wave with a decrease to $A_{max}/\sqrt{2}$.

For the coupling factor $k$, the following holds true:

$$k = \frac{R_p}{Z_L} \quad \text{at the parallel-resonant circuit} \tag{3.10}$$

and

$$k = \frac{Z_L}{R_r} \quad \text{at the series-resonant circuit} \tag{3.11}$$

The coupling factor specifies which portion of the energy received by the resonator is used in the outer charge. We designate

$k = 1$, optimum coupling;
$k < 1$, loose coupling;
$k > 1$, tight coupling.

The input impedance of the resonator feeder cable line is dependent upon its length $l$ as a result of its transformation behavior. Figure 3.5c shows the curves of the input impedance for $l = 0$ and $l = \lambda_r/4$. The input reflectivity becomes minimal at the resonant frequency $f_r$. For the resonant standing-wave ratio, the following is valid:

for (3.12)

$$s = \frac{Z_L}{R} \quad \text{for} \quad R < Z_L$$

and

$$s = \frac{R}{Z_L} \quad \text{for} \quad R > Z_L\,. \tag{3.13}$$

The *tuning* of cavity resonators can follow from the alteration of the cable length with a shorting plunger. The shorting plunger must only touch the walls at those places where the axial wall current paths from the resonator walls to the plunger must be closed. A change in the resonant frequency can also follow by introducing conductors (diaphragms or pins) in the resonator. When this conductor displaces the electrical fields (capacitative reactance), the resonant frequency is lowered; if it displaces the magnetic fields (inductive reactance), then the resonant frequency is raised. Tuning is also possible through the introduction of dielectric material. Electronic tuning can follow with the help of variable reactor diodes (see *Mikrowellentechnik,* Vol. 2, Ch. 3) or polarized ferrites (e.g., YIG components, Ch. 5).

Next, we will examine the various field modes in rectangular *cavity resonators:*

For the resonant wavelength, we obtain with Eq. (3.4) by inserting the critical wavelength $\lambda_c$ for rectangular hollow waveguides from Eq. (1.165):

$$\lambda_r = \frac{c}{f_r} = \frac{2}{\sqrt{\left(\frac{m}{a}\right)^2 + \left(\frac{n}{b}\right)^2 + \left(\frac{p}{c}\right)^2}} \, . \tag{3.14}$$

The form of the resonator fields is characterized by adding the value $p$ as a third index to the code number of the stimulating field mode (e.g., $H_{101}$ resonance). Here the third index $p$ specifies the number of half-cycles in the field distributions in the z-direction.

With $E$-waves, resonator performance is possible with $p = 0$ ($\lambda_h \rightarrow \infty$). We then obtain resonator length $\lambda_r = \lambda_c$. $E_z$ becomes independent of $z$ and $E_x = E_y = 0$. These modes with $p = 0$ are not possible for $H$-waves, since with $H$-waves the electrical field is strictly directed transversely, and in order to meet the limit condition $E_{tan} = 0$ at the separations, a very high value is necessary. With $E$-waves the electrical cross-components disappear for $\lambda_0 = \lambda_c$ (for example, see $E_{110}$ resonance in Fig. 3.8).

We shall presently investigate the field distribution for the $H_{10}$-mode in a rectangular cavity resonator. The superposition of incident and reflected wave yields the $E_y$ components:

$$E_y = \hat{E}_h \cdot \sin \frac{\pi x}{a} \cdot \cos(\omega t - \beta_h \cdot z) + \hat{E}_r \cdot \sin \frac{\pi x}{a} \cdot \cos(\omega t + \beta_h \cdot z) \, . \tag{3.15}$$

For the short-circuit hollow waveguide, the following is valid:

$$\hat{E}_h = -\hat{E}_r \, . \tag{3.16}$$

If we examine the relationships along the median plane $x = a/2$, then we obtain with Eq. (3.15)

$$E_y = \hat{E}_h \cdot [\cos(\omega t - \beta_h \cdot z) - \cos(\omega t + \beta_h \cdot z)] = 2\hat{E}_h \cdot \sin\beta_h \cdot z \cdot \sin\omega t \,. \tag{3.17}$$

The magnetic lines of force envelop the electrical shift current density distribution, for which the following is valid:

$$\vec{S}_V = \frac{d\vec{D}}{dt} \,. \tag{3.18}$$

This is

$$S_{Vy} = \varepsilon_0 \cdot \frac{dE_y}{dt} = \varepsilon_0 \cdot \omega \cdot \hat{E}_h \cdot [-\sin(\omega t - \beta_h \cdot z) + \sin(\omega t + \beta_h \cdot z)]$$

$$= \hat{S}_{Vy} \cdot [-\sin(\omega t - \beta_h \cdot z)] + \sin(\omega t + \beta_h \cdot z)] \,. \tag{3.19}$$

We obtain the following relations:

$$\begin{aligned}
t &= 0 &&: E_y = 0, && S_{Vy} = 2S_{Vy} \cdot \sin(\beta_h \cdot z) \\
t &= T/4 &&: E_y = 2\hat{E}_h \cdot \sin(\beta_h \cdot z), && S_{Vy} = 0 \\
t &= T/2 &&: E_y = 0, && S_{Vy} = -2S_{Vy} \cdot \sin(\beta_h \cdot z) \\
t &= 3T/4 &&: E_y = -2\hat{E}_h \cdot \sin(\beta_h \cdot z), && S_{Vy} = 0 \,.
\end{aligned} \tag{3.20}$$

The time shift of the electrical shift current density by 90° with respect to the electrical field intensity does not signify a local displacement as is the case for incident waves in the hollow waveguide by $\lambda_h/4$ in the direction of propagation. Whereas with incident waves the electrical and magnetic fields oscillate in phase, here the electrical and magnetic fields, which are in phase with the shift current density, are in quadrature phase delay. The energy oscillates between the electrical and magnetic fields in a fashion similar to the case of a resonant circuit with concentrated circuit components. In Fig. 3.6, the field and wall current distribution are outlined in the case where both reflecting walls are located at a distance of $\Delta z = \lambda_h/2$ from each other; therefore, for the $H_{101}$ resonance.

In the case of $E_{110}$ resonance, we obtain an analogous path of the lines of force, whereby the magnetic lines of force now run parallel to the $xy$-plane and the electrical lines of force run perpendicular to it.

We will next examine a *circular cylindrical cavity resonator:*

The resonator wavelengths of circular cylindrical cavity resonators can be calculated according to Eq. (3.4), whereby the critical wavelengths of the circular hollow waveguide for $E$-waves can be calculated according to Eq. (1.197) and for $H$-waves according to Eq. (1.205), or they may be taken from Table 1.1.

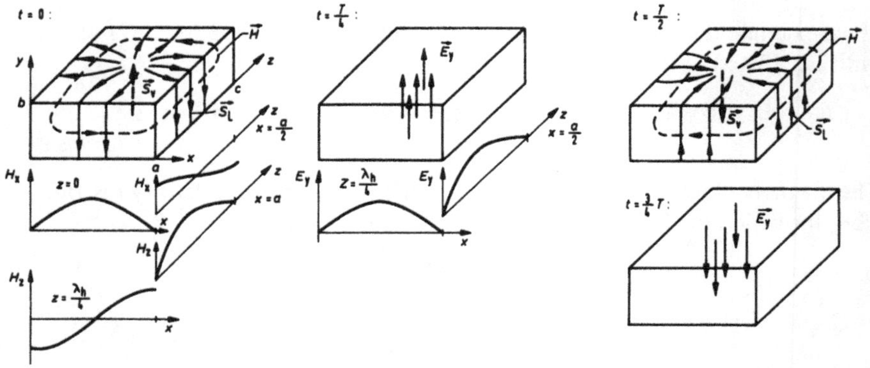

**Fig. 3.6** Rectangular cavity resonator with field and amplitude distribution and wall currents of the $H_{101}$-resonance

In order to characterize the resonance, we add the index $p$, which specifies the number of half-cycles of the field distribution in the $z$-direction, and to the code number of the respective field mode (e.g., $E_{010}$ resonance).

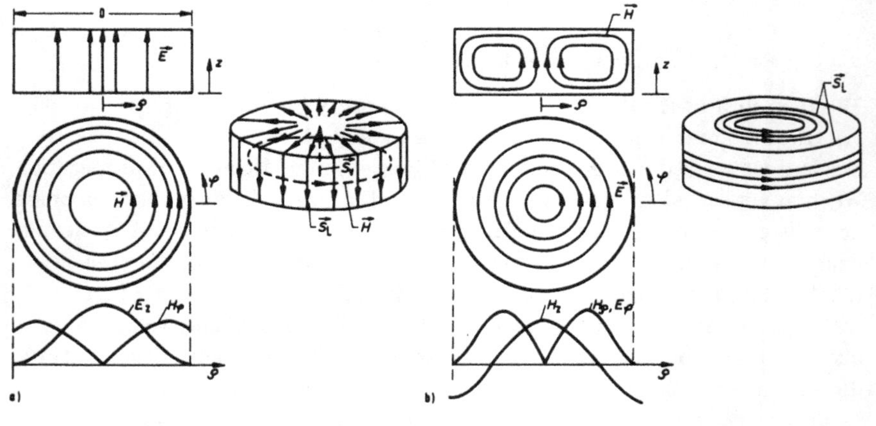

**Fig. 3.7** Circular cylindrical cavity resonator with field and amplitude distribution and distribution of the wall currents: (a) for the $E_{010}$-resonance, (b) for the $H_{011}$-resonance

Figure 3.7 shows the field and wall current distribution for the $E_{010}$ and $H_{011}$ resonance. With $H_{0\,np}$ resonances of the circular cylindrical cavity resonators, no wall currents exist in the $z$-direction.

As in the case of cavity resonators with a rectangular cross section, with $E$-resonances we also obtain the special case $p = 0$ with $\lambda_r = \lambda_c$. The field components are, therefore, independent of $z$. The electrical lines of force run parallel to the cylinder axis. Figure 3.8 shows, as an example, the field distribution for an $E_{010}$ circular cylindrical resonator.

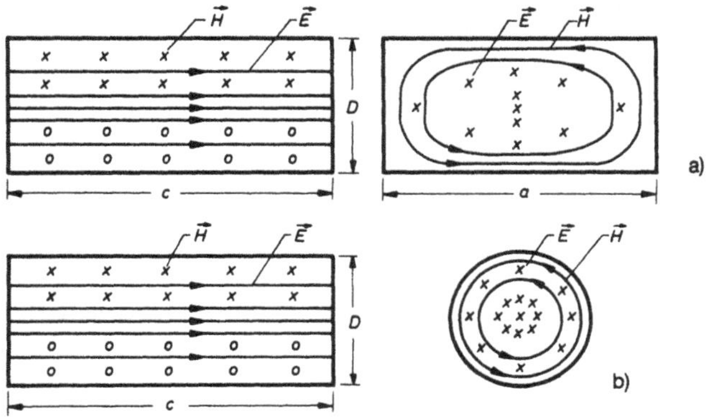


**Fig. 3.8** Fields in cavity resonators with $\lambda_r$ independent of C: (a) $E_{110}$-rectangular resonator, (b) $E_{010}$-circular cylinder resonator

We designate self-oscillation modes as *abnormal* (or *degenerate*) when they display the same resonant frequencies at different field distributions.

Tunable resonators can be used as *wave meters* in order to measure frequency. In this connection, they can be constructed as a single-gate and connected as the terminal element to a line or laterally coupled to a transit line. Therefore, in the case of resonance they function reflectively (reflection wave meter) or they absorb the maximum active power as an absorption wave meter in the case of resonance (absorption mode). They can also be constructed as a dual-gate and are connected as a transmission frequency meter (transmission mode) between the generator and the receiver, whereby in the case of resonance they transfer the maximum active power to the receiver. With a direct-indicating wave meter, the small portion of energy decoupled by resonant tuning is used directly for an indicator (as a detector and reading instrument). With the absorption wave meter, the resonance is determined by the power drop in the secondary line.

The $Q$-factor and, therefore, the frequency resolution become higher when the resonant length of the cavity resonator amounts to many half-wavelengths. We obtain an improvement in the sensitivity of the frequency measurement by evaluating the rapid phase change in the vicinity of the resonator. To this end, we can use a hybrid-T-bypass (sec. 3.7), in which the source is coupled to the parallel arm (Fig. 3.32) and a detector is coupled to the lateral arm. The wave meter is attached at the arm of the main line and a shorting plunger is attached at the other arm whose shift is a criterion for frequency in the realization of the measuring alignment.

The use of resonators in filter circuits will be discussed in sec. 3.4.

## 3.2 REACTANCES (REFLECTORS)

### 3.2.1 Dummy Lines

As was shown in sec. 1.5.2, reactances can be realized with open- or short-circuited terminal components. Such a lateral short-circuit (idling) dummy line, which is connected to a cable and causes a reactance, is also known as a *stub line.*

In this manner, a waveguide component that is short circuited at the terminal can be laterally connected as a dummy line to a hollow waveguide. If the vertical currents of the transit line are suppressed by the resulting aperture, then we obtain a series reactance; if the cross-currents are suppressed, then we obtain a parallel conductance (see diagram in Fig. 3.9). The coupling of the lateral hollow waveguide can follow from a slot as well. A lateral coaxial dummy line can also be used, whereby its inner conductor reaches across the hollow waveguide.

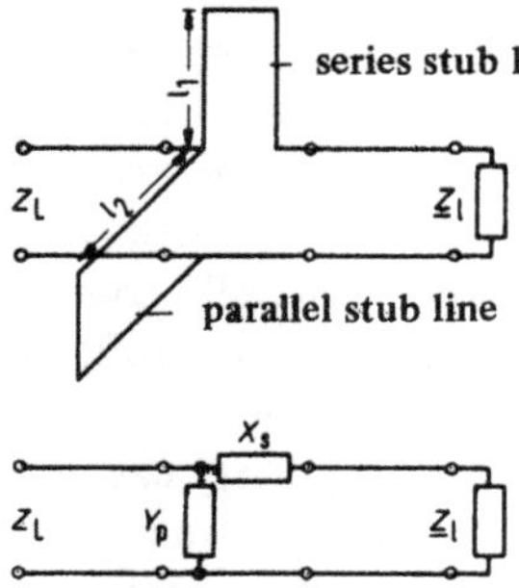


**Fig. 3.9** Parallel- and series-stub lines with equivalent circuit

The reactance caused by a dummy line can be varied by changing its length with a shorting plunger. With such an adjustable reactance, which is realized with a variable short circuit, (theoretically) all reactances $-\infty < X < \infty$ can be adjusted by a mechanical shift of the short-circuit plane. In order to realize a specified short circuit unimpaired by transfer resistances caused by mechanical contact between the plunger and the circuit wall, we create the short circuit by a transformation with a circuit component on the front side of a movable plunger which is immersed in the line (and deals with the mechanical contact at a location of minimal current). Figure 3.10 shows such a variable short circuit: the rear short-circuit plane (at C) is transformed into an idling plane at B over the middle $\lambda/4$ circuit component, and on a further $\lambda/4$ line into a short circuit at A.

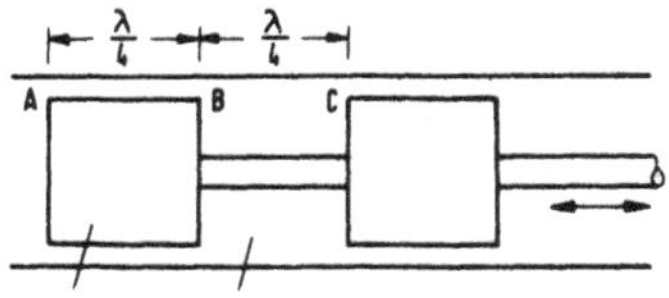


**Fig. 3.10** Variable short circuit (shorting plunger)

### 3.2.2 Diaphragms and Pins

Reactances which, for example, are necessary for the construction of filters as well as for purposes of tuning and adapting can be realized by mounting diaphragms or pins in the waveguide. If we bring such an obstacle into the circuit, then in its vicinity the original field will be disturbed while meeting the altered limit conditions. At the obstacle, reflected waves will be triggered that propagate in both directions, and new field modes (attenuation modes) are stimulated. These attenuation modes are not capable of propagating and die down within a short distance from the center of disturbance. If we assume that the obstacle is an ideal conductor, then it operates entirely reflectively. This means an accumulation of energy and can be represented in an equivalent circuit by a reactance or a conductance. In either case, whether the electrical or the magnetic field energy received predominates at the field excited, the disturbance behaves as a capacitative or an inductive conductance. If the energy received in the electrical and magnetic fields is equal, then resonance occurs.

The characteristic values of the concentrated structural components of an equivalent circuit, which describe the effect of such a center of disturbance (in a given frequency range), are also a function of frequency and the transmitted wave mode as well as the geometry of the center of disturbance.

We will first examine the effect of a cross-section constriction of an $H_{10}$-wave at some point on a rectangular hollow waveguide through an all-round metal diaphragm according to Fig. 3.11. In this representation, the path of the electrical lines of force, or the shift current lines and the wall currents are outlined. In the diaphragm, additional wall currents are excited and, therefore, the cross-current paths are lessened. This yields an inductive effect which can be (approximately) represented in an equivalent circuit by a shunt inductance. The electrical lines of force are reduced and, therefore, the shift currents forming the continuation of the wall currents are increased. This signifies a capacitative component which can be described (approximately) by a shunt capacity. Thus, we obtain a parallel resonant circuit as an equivalent circuit for the all-round metal diaphragm. At the resonant frequency, the shunt conductance becomes zero; the diaphragm remains practically without effect and energy transmission can take place in the hollow waveguide. Once outside the resonant frequency, the obstacle has a reflecting effect; for lower frequencies one obtains an inductive effect and for higher frequencies a capacitative effect.

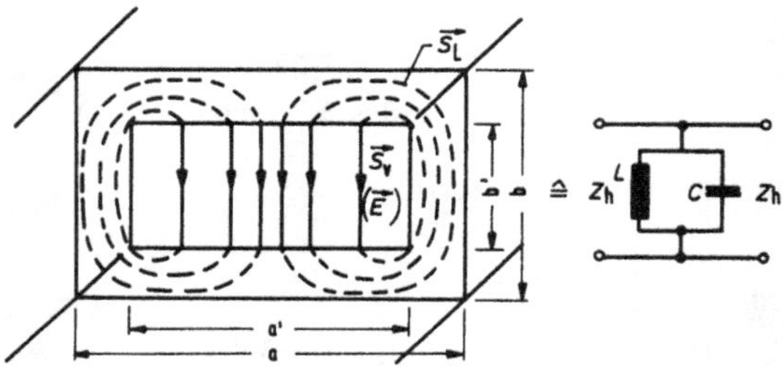


**Fig. 3.11** All-round diaphragm in an $H_{10}$ rectangular hollow waveguide and simple equivalent circuit

For small diaphragm thickness, we obtain as an approximation formula [10] for the resonant wavelength ($Y = 0$)

$$\lambda_r = 2a \cdot \sqrt{\frac{\left(\frac{a'}{a} \cdot \frac{b}{b'}\right)^2 - 1}{\left(\frac{b}{b'}\right)^2 - 1}} . \tag{3.21}$$

If the diaphragm is formed such that the effect of the reduction of the electrical lines of force predominates, then it has a capacitative effect. With a purely capacitative diaphragm (Fig. 3.15e) $\lambda_r = 2a$ with $a' = a$. The concentration of the electrical field energy can be (approximately) represented by a capacitative parallel conductance. The degree of reflection caused in the line by the dummy element increases with a higher frequency.

If the influence of the lateral localization of the waveguide cross section is stronger, then we obtain an inductive diaphragm effect. The resonant frequency of the purely inductive diaphragm ($b' = b$, Fig. 3.15a) lies above the waveguide's operating range; therefore, the shunt conductance is always inductive. Convection currents produce an additional magnetic field flow in the thin lateral metal strips as well as in the continuous bar, according to Fig. 3.15b, because of the potential difference. As a result of this accumulation of magnetic field energy, we obtain (approximately) the effect of an inductive parallel conductance. The degree of reflection produced in the line by this method decreases with a higher frequency.

With a capacitative diaphragm, the additional capacitance increases with a decreasing diaphragm aperture and an increasing diaphragm thickness. With an inductive diaphragm, the additional inductivity decreases with an ever decreasing diaphragm aperture and with an increasing diaphragm thickness.

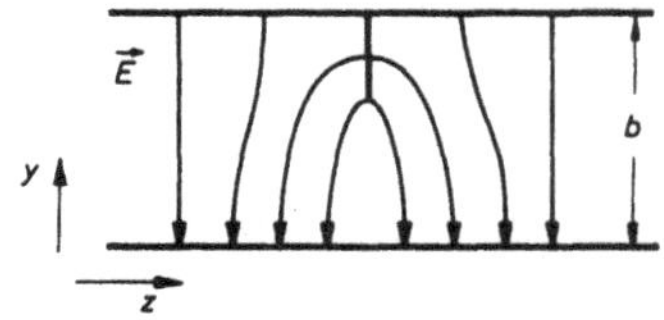


**Fig. 3.12** Electrical field in the vicinity of a capacitative diaphragm

Figure 3.12 shows the change in the electrical field in the vicinity of a capacitative diaphragm in a section parallel to the $yz$-plane. The electrical field intensity receives a component in the vertical direction, which signifies the excitation of additional $E$-fields which are not capable of propagation. Figure 3.13 shows how the magnetic field is influenced by an inductive diaphragm. Here, $H_{m0}$-fields which are not capable of propagation are added to the $H_{10}$-wave.

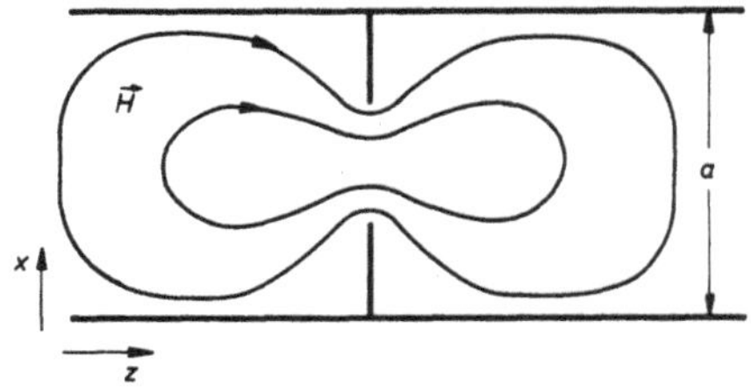


**Fig. 3.13** Magnetic field in the vicinity of an inductive diaphragm

The cylindrically conducting *pin* is also often used as a dummy element. In Fig. 3.14, the influence of such a pin, which is located parallel to the narrow side of the rectangular hollow waveguide, upon the propagation of an $H_{10}$-wave is examined. The electrical lines of force of the hollow waveguide produce line currents in the pin which continue to its terminal as shift currents in the opposite wall. The effect of this line current with its magnetic field can be (approximately) described by an inductance, and the effect of the shift current by a capacitance located along a row for this purpose. With a short pin length immersion depth $s$, the shift currents in the hollow waveguide are increased and the pin has a capacitative effect, whereby the susceptible capacitance becomes larger with an increasing pin thickness. With a longer pin length, the influence of the pin's magnetic field predominates and, thereby, the inductive component as well. There is an intermediate pin length for which the capacitative and inductive effects are equally large: the pin shows a series resonance; its effect is that of a shunt conductance tending toward infinity (short circuit). The resonant wavelength $\lambda_r$ of the pin increases with its length $s$. For thin pins, the following is valid

$$\lambda_r \approx 4s \tag{3.22}$$

(resonance at $s/b \approx 0.7$-$0.9$). With an increasing pin thickness, $\lambda_r$ decreases. For $s = b$, then $\lambda_r = \lambda_c = 2a$. At the series resonant frequency $f_r$, the energy in the hollow waveguide is reflected and outside the resonant frequency transmission takes place. For $f < f_r$ the relative conductance (with respect to the reciprocal oscillation resistance of the hollow waveguide) is $B_{rel} > 0$, which means capacitative; and for $f > f_r$ it is $B_{rel} < 0$, therefore, inductive. The tuning range attainable with a pin is smaller than with a dummy line.

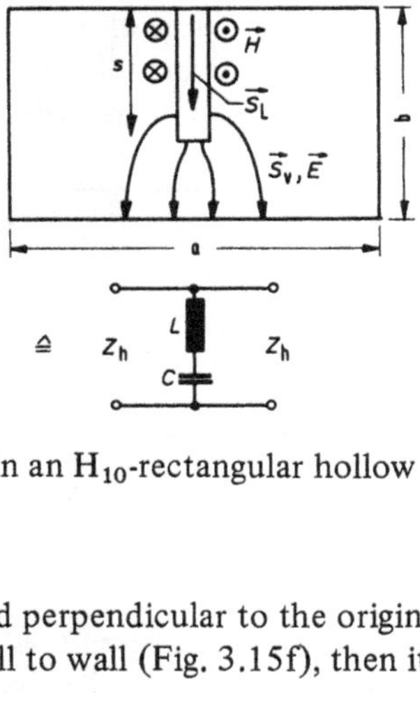


**Fig. 3.14** Conducting pin in an $H_{10}$-rectangular hollow waveguide and simple equivalent circuit

If a metal pin is arranged perpendicular to the original electrical field (parallel to the broad side) from wall to wall (Fig. 3.15f), then it functions as a susceptible shunt capacitance.

Dielectric pins or plates are also used as transformation elements and have a capacitative effect because of the increase in the shift currents.

If the hollow waveguide is not operated by its surface wave, then the obstacle is to be constructed such that no other field modes arise which are capable of propagation.

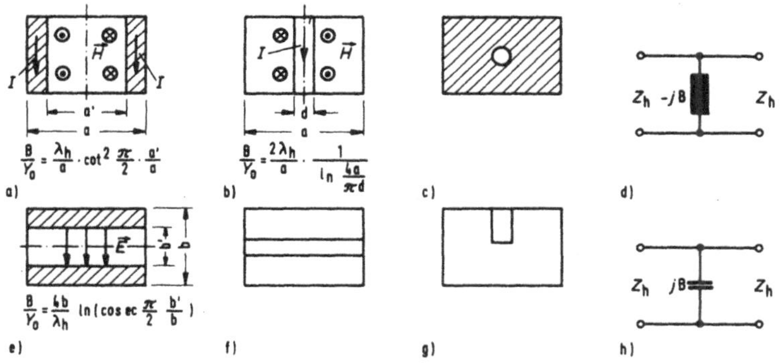


**Fig. 3.15** Dummy circuit in an $H_{10}$-rectangular hollow waveguide with approximation formulas: (a-d) capacitative parallel conductances, (e-h) inductive parallel conductances

In Fig. 3.15 some dummy circuit components are assembled which function as capacitative or inductive parallel conductances for a rectangular hollow waveguide with $H_{10}$-wave propagation. Figure 3.16 shows various diaphragm forms for an $H_{11}$ circular hollow waveguide. Some approximation formulas (according to [3]) are specified for the relative susceptance of the components.

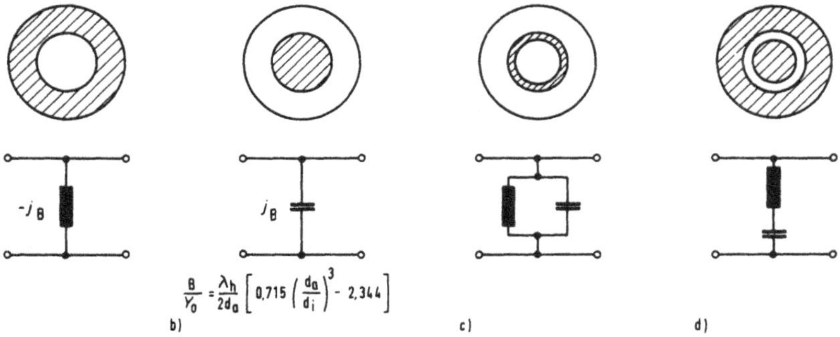


**Fig. 3.16** Thin metal conductor in an $H_{11}$-circular hollow waveguide

Applications of the reactance described in the form of dummy lines, diaphragms, or pins can be obtained by accomodating connections and filter circuits among others.

## 3.3 ACCOMMODATING CONNECTIONS

We are often faced with the task of accommodating an impedance to the line oscillation resistance. This is possible with the help of structural components which function as dummy elements, or through the use of transformation characteristics of circuit components.

Such a reactive component necessary for removing a mismatching can be realized with the help of a *dummy line,* or a *tie line,* which is laterally connected to the circuit (sec. 3.2.1). Figure 3.17 shows the diagram of a parallel and a series tie line for the matching of $\underline{Y}_A$ or $\underline{Z}_A$ to a line. The execution of a circuit bypass in waveguide engineering as a parallel divider (*H*-divider), or as a series divider (*E*-divider), is represented in Fig. 3.29. With an arrangement of three tie lines, all different terminal impedances on a line can be adapted. A practical aid in the construction of tie lines is the Smith Chart (Ch. 1).

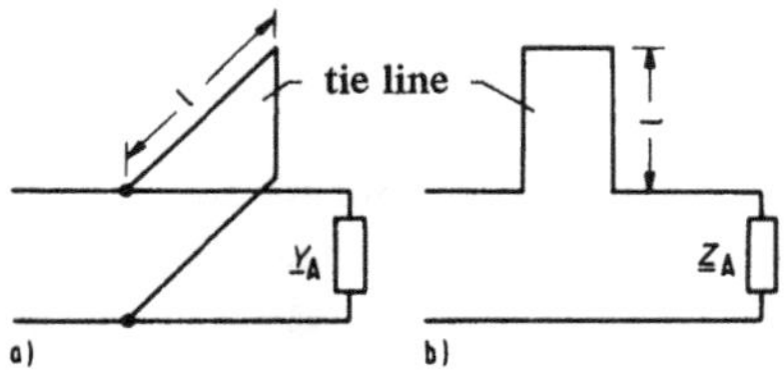


**Fig. 3.17** (a) parallel-tie line and (b) series-tie line for matching

We often use *diaphragms* and *pins* (sec. 3.2.2) as dummy elements for a matching transformation on hollow waveguides in the conversion of the mismatching. In the case of a *slide-screw transformer,* we use a pin which (similar to the case of a circuit for sound- and flash-ranging station) is displaceable (to the phase change) in a narrow vertical cross section in the center of the waveguide breadth, and whose immersion depth is adjustable in the hollow waveguide. Simpler is an arrangement of two screws at a distance of $\lambda_h/4$ from each other in the center of the waveguide breadth, whose immersion depth is adjustable in the hollow waveguide; or the triple-screw transformer with three screws located at an equal distance (approximately 3/8 $\lambda_h$) from each other, with which every mismatch can be compensated by adjusting the immersion depth of the screws.

Movable dielectric elements (pins) are also used in coaxial transmission lines and hollow waveguides as variable shunt capacitors for accommodation purposes.

Very often we use a $\lambda/4$ line for accommodation purposes, which is also designated as a *$\lambda/4$ transformer* (Fig. 3.18). A $\lambda/4$ line with oscillation resistance $\underline{Z}_L$ transforms, according to Eq. (1.116), a terminal resistance as

$$\underline{Z}_1 = \frac{\underline{Z}_L^2}{\underline{Z}_2} \tag{3.23}$$

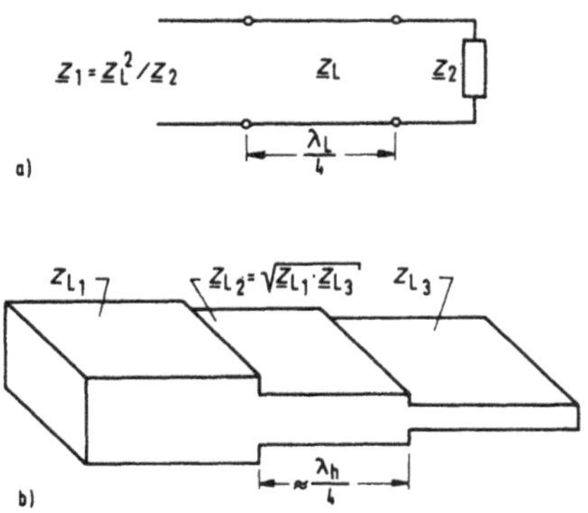


**Fig. 3.18** λ/4-transformer: (a) general, (b) in the execution of a rectangular hollow waveguide

at the input. The λ/4 transformer is used as an intermediate line for matching two lines with different line resistances, or as a matching element for structural circuit components. In this manner, for example, in the arrangement according to Fig. 3.18, a diode can be mounted in the waveguide with the lower oscillation resistance $Z_{L_3}$ in the $E$-field direction, and its low ohmic impedance can be adapted to the high waveguide oscillation resistance $Z_{L_1}$.

Accommodation with the help of simple diaphragms, pins, or λ/4 lines is of relatively narrow bandwidth.

With *wideband accommodation,* a matching element may be used which, in addition to suitable susceptance, also has suitable frequency dependence; for example, a diaphragm with a resonance in the transmission range.

A wider band accommodation than in the case of a single-stage λ/4 transformer is realized with multiple-stage λ/4 transformers through the use of several λ/4 line cross sections which are connected in tandem and graded in ordered formation by oscillation resistance. An example of filter characterized by maximum flatness is the binomial transformer with a step increase of the oscillation resistances according to binomial coefficients, and for a larger filter bandpass width one obtains a uniform standing-wave ratio with the Chebyschev transformer.

Wideband accommodation is also possible with the help of *inhomogeneous lines,* for example, in the form of a line with a change of cross-section area which runs linearly (taper, Fig. 3.25), or as a transmission running exponentially (exponential line) between the lines of different cross sections to be adapted.

The use of the magic $T$ in the form of the $E$-$H$ tuner as an accommodating connection will be described in sec. 3.7.

## 3.4 FILTER CIRCUITS

In the design of microwave filters, we can proceed from the corresponding filter circuits with concentrated capacitors and inductors or resonant circuits, and realize these through circuit cross sections, for example, as dummy lines laterally connected to the hollow waveguide or as microwave resonators. In the same manner, the necessary reactances may be realized with the help of diaphragms and pins. To this end, in the use of hollow waveguides, their high-pass behavior can be exploited.

Resonators are often used in the construction of filters. These can be designed as transit filters (transmission filters) or as suppression filters (reflection filters). Therefore, with a *suppression filter,* we can suppress a path junction for a given frequency range by coupling a resonator, which represents, as its resonant frequency, a very high resistance lying along a row with respect to the line; or a very low resistance connected in parallel. A through-connecting extension with a *transit filter* for a given frequency range can be obtained by coupling the incoming and outgoing lines to a dual-gate resonator, which then creates the connection at its resonance over the excited fields. The more secure the coupling at the resonator, the wider the band of the transit behavior becomes. Figure 3.19 shows the diagram of a waveguide transit filter with one circle, at which the resonator area is coupled over the diaphragms. In an equivalent circuit with concentrated circuit components, the filter can be represented by a series-resonant circuit and two coupling transmitters. The resistance $R$ represents the filter's losses (surface resistance). Additionally, the load resistance $R_a = Z_L$ and the internal generator resistance $R_i$ are transformed over the transmitter with the transformation ratio $u$ in the resonating circuit. Thereby we obtain for the resonance $Q$-factor of the loaded resonant circuit

$$Q = \frac{f_r}{\Delta f} = \frac{\omega_r \cdot L}{R + \frac{1}{\ddot{u}^2}(R_i + R_a)} . \tag{3.24}$$

With several coupled resonators in multiple-circuit filters, the filter characteristics can be improved (steeper flanks) and larger bandwidths can be produced. With equal resonators a Chebyschev* filter can be constructed. It has very steep sides, but in the transmission range it possesses a standing-wave ratio whose magnitude increases with the number of resonators. For a filter with maximum flatness, or Butterworth filter, the standing-wave ratio is eliminated in the transmission range by allowing all the roots of the transmission function to coincide. Filter tuning can follow by way of a tuning screw which has a capacitative effect, or tuning can take place electronically with magnetized ferrites (e.g., YIG-elements).

*C. Rint, ed., *op. cit.,* Vol. 1, pp. 105 ff, 124 ff.

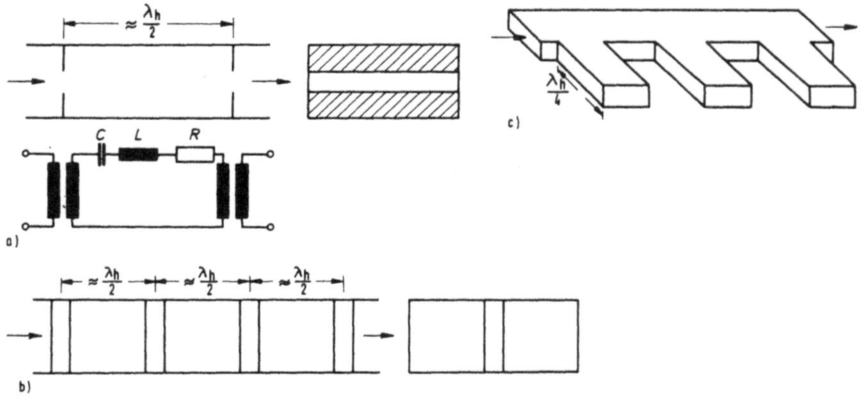


**Fig. 3.19** (a) waveguide transit filter with one circle with equivalent circuit, (b) three-circuit filter, (c) comb filter

In Fig. 3.19b the diagram of a three-circuit filter is shown, for which the coupling elements are constructed as pins. Figure 3.19c shows a so-called *comb filter* in waveguide construction. To this end, short-circuit cavities (tie lines) which function as resonant circuits are connected to a hollow waveguide.

Occasionally, in multiple-stage filters, we may take advantage of the possibility of creating the effect of coupled resonators through a single resonator in which various field modes are present. Through small inhomogeneities, such as metal pins or loops, at suitable locations of the resonator an energy exchange is forced between the propagating modes from the input of the resonator to its output.

## 3.5 EFFECTIVE RESISTANCES (ABSORBERS)

In the case of effective resistances, terminal resistances and (transit) attenuator pads can be differentiated.

Non-reflecting *terminal resistances* are also designated as an absorber or wave basin (noted by the switch symbol in Fig. 3.20c). A simple way to achieve low-reflection terminal resistance in a hollow waveguide is offered by a resistance supporting film mounted at a distance of approximately $\lambda_h/4$ diagonally in front of the short-circuit end of the waveguide. The supporting film is mostly comprised of non-conducting column supports (ceramic or glass) which are coated with coal or metal. In the surface resistance layer, which is mounted parallel to the original electrical field, convection currents flow as a result of the electrical field and, therefore, the electromagnetic energy of the wave is transformed into heat. Such a termination is, however, strongly dependent on frequency. A wideband termination can be realized by a gradual increase in attenuation whereby we mount, according to Fig. 3.20a, a supporting film with an increasing height in the horizontal direction of the hollow waveguide. For better performance, we may realize connected-circuit terminations, according to Fig. 3.20b, such that the waveguide end is provided with a wedge (that is also repeatedly beveled) made of a sharply attenuated material (dielectric material mixed with HF-iron dust, or a mixture of graphite and sand, or ferrite; and also possibly wood). The heat dissipation can be improved by cooling vanes or circulating water.

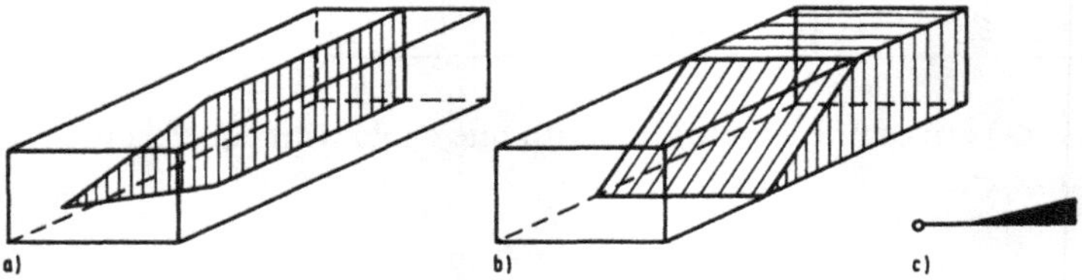


**Fig. 3.20** Terminal resistance: (a) for low and (b) for high performance, (c) switch symbol

Adjustable *attenuator pads* can be produced by introducing vertically directed small plates furnished with an absorbing layer in a hollow waveguide. In the type model, according to Fig. 3.21a, an attenuation supporting film is rotated in the low-frequency "bread knife" form (maximum length approximately $2\lambda_h$) in an elongated slot in the waveguide's broad side; therefore, at a location of maximum electrical field intensity. The supporting film is made of a poorly conducting graphite paper or a metallically loaded mica. With the so-called *vane* or *strip attenuator* (Fig. 3.21b), the resistance vanes are adjustably arranged in the horizontal direction. The attenuation is strongest on the $H_{10}$-wave in the center of the rectangular hollow waveguide since the electrical field is maximal there. With the attenuation adjustment, the electrical length (phase) also changes; additionally, the attenuation is frequency dependent.

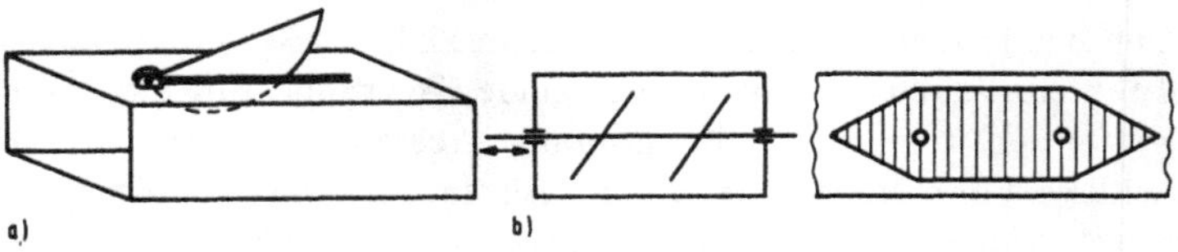


**Fig. 3.21** (a) "breadknife"-attenuator, (b) vane- or strip-attenuator

In addition to the absorbent attenuator pads, reflecting attenuator pads are also used. With these so-called *hollow tube voltage dividers,* we take advantage of the exponentially suppressed, aperiodic field propagation of a (circular) hollow waveguide operated below its critical frequency.

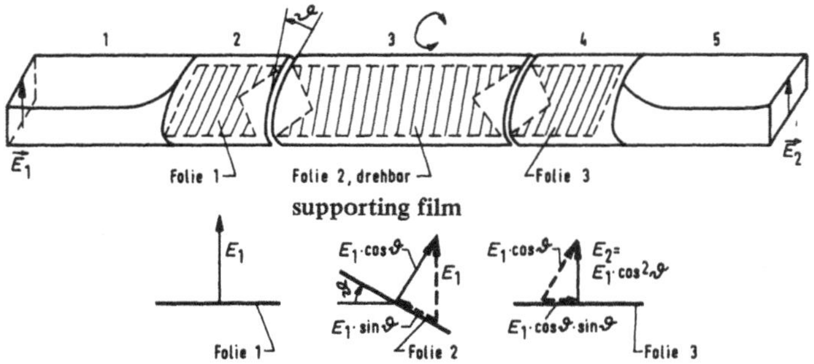


**Fig. 3.22** Rotating supporting-film attenuator pad, precision attenuator

The reflections of a reflection attenuator pad are often disturbing. For transmission efficiency, pads are avoided. Such an attenuator pad, which also avoids the additional phase change in the case of variable attenuation, is the *rotating supporting-film attenuator pad,* which is also known as the precision attenuator (Fig. 3.22). The signal approaching as an $H_{10}$-wave in a rectangular hollow waveguide is transformed by a wave mode transformer (1) into an $H_{11}$-wave in a circular hollow waveguide. The variable attenuation of the wave is adjusted by a rotatable resistance supporting film, and, finally, is once again converted by a transition from the circular to the rectangular waveguide cross section into an $H_{10}$-wave. The wave mode filter (2 and 4), with the fixed absorbing vanes parallel to the broad side of the rectangular hollow waveguide, should suppress those wave segments that are polarized perpendicular to the direction of polarization in the rectangular waveguide. The middle part of the attenuator pad is made up of a rotatable $H_{11}$-circular waveguide segment in which a thin, poorly conducting resistance supporting film (e.g., with a sheet mica attenuated with NiCr-alloy) is mounted inside a diameter plane. At this small plate the segment of the $H_{11}$-wave whose electrical field is directed parallel to the small plate is absorbed, whereas the segment which is parallel to it remains practically unaffected. If the supporting film is parallel to the electrical field, then we obtain maximum attenuation, and in a perpendicular position the attenuation is minimal. The incident $H_{10}$-wave possesses an electrical field intensity $E_1$, and the mid-section should be rotated at the angle $\theta$ with regard to the direction of least attenuation. The partial wave with the electrical field intensity component $E_1 \cdot \sin\theta$, which lies parallel to the rotatable vane, would be (approximately) completely absorbed, and the partial wave with the field component $E_1 \cdot \cos\theta$, which is perpendicular to it, is not attenuated at all. In the self-connecting waveguide cross section, the fixed attenuation supporting film (3) of the partial wave with the field component $E_1 \cdot \cos\theta \cdot \sin\theta$ is completely absorbed, and the partial wave with $E_2 = E_1 \cdot \cos\theta \cdot \sin\theta$ is allowed to pass. Therefore, we obtain for the attenuation of the rotating supporting-film attenuator pad as a function of the rotation angle $\theta$

$$\frac{a}{\mathrm{dB}} = 20 \log \frac{E_1}{E_2} = 40 \log \left(\frac{1}{\cos\vartheta}\right). \tag{3.25}$$

This attenuator pad has the advantage that the attenuation value within the applicable frequency range of the hollow waveguide (1:1.5) is not dependent on frequency and the phase constant is independent of the adjusted attenuation.

Attenuator pads are often realized by way of PIN-diodes (see *Mikrowellentechnik,* Vol. 2, Ch. 3) or ferrites (Ch. 5).

Absorbing adapted circuit terminations are also manufactured as structural components, for which the absorbed charge is used as a *power measurement* or a signal reading. For power measurements under approximately 1 W, temperature-dependent heating grids are used, such as ballast resistors (bolometer, thin metal wire, or thin conducting layer with positive temperature coefficients) or thermistors (small semiconductor pill with negative TK). We arrange these components at a location of maximum electrical field intensity of approximately $\lambda/4$ in front of the circuit terminal in the cross section of the conducting wire. At the component, an absorption of microwave power takes place and the heating of the component due to the high-frequency current calls for a variable resistance which can be measured with a low-frequency measuring bridge. With matching links, we can obtain a wave which is almost completely absorbed in the measuring head-end.

If an absolute power measurement is not necessary, then a semiconductor rectifier diode can also be used as a detector. In order to measure higher power, we use a dry or liquid calorimeter.

## 3.6 PHASE SHIFTERS

Just as in the case of attenuator pads, there are many types of microwave phase shifters, of which a few will be briefly described.

*Hollow waveguides partially filled with dielectric material* (specific inductive capacity $\epsilon_r = \epsilon' - j\epsilon''$) are used among other things as phase shifters ($\epsilon'' = 0$) or attenuator pads ($\epsilon'' \neq 0$, $\epsilon' < \epsilon''$).

In the rectangular hollow waveguide with non-uniform dielectric, generally, combined $E$- and $H$-waves occur as so-called *longitudinal waves.* If the electrical field of these waves only has components in the longitudinal plane through the hollow waveguide (parallel to the *xz*-plane), then we speak of *E*-longitudinal waves (*LSE*-field); if the magnetic field intensity only has components in a longitudinal plane (*yz*-plane), then we are dealing with a so-called *H*-longitudinal wave (*LSH*-field). Figure 3.23 shows the superposition of the field components of *E*- and *H*-waves with those of longitudinal waves.

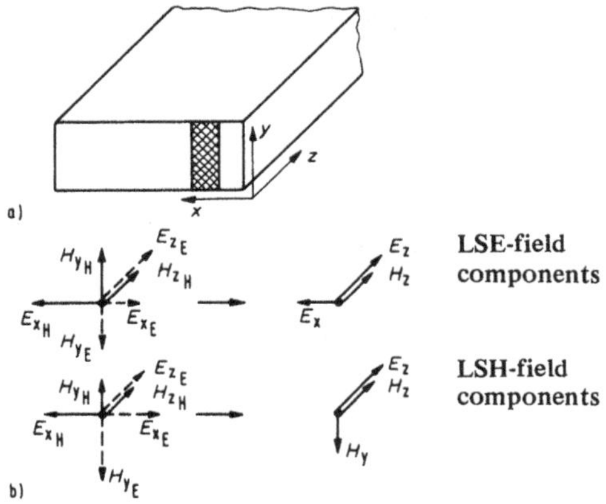


**Fig. 3.23** Superposition of E- and H-waves with longitudinal waves

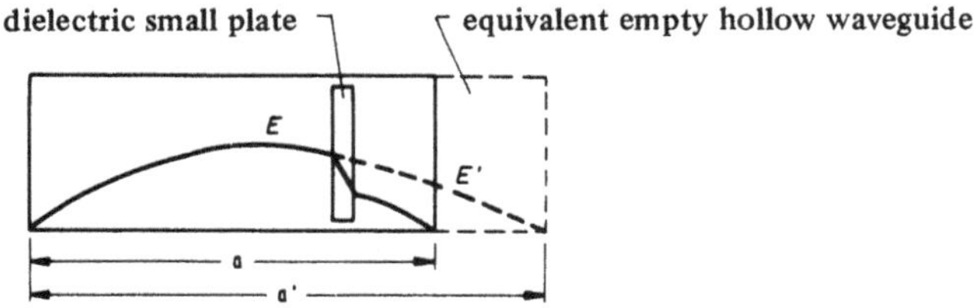


**Fig. 3.24** Dielectric phase shifter

A phase shifter constructed similarly to the vane attenuator works with a small plate (or bar) made of a low-loss dielectric material in the vertical (longitudinal) direction of an $H_{10}$ rectangular hollow waveguide. As Fig. 3.24 shows, with the distribution of the electrical field disturbed by the small plate, the small plate has the effect of widening the hollow waveguide, and thereby increasing the critical wavelength ($\lambda_c = 2a$). With this the waveguide wavelength

$$\lambda_h = \frac{\lambda_0}{\sqrt{1 - (\lambda_0/\lambda_c)^2}}$$

is decreased, and the phase with respect to the hollow waveguide is changed. The influence is greatest in an area of high electrical field intensity at the center of the waveguide. The small plate may be introduced through an elongated slot in the middle of the waveguide's broad side, or can be slid through the narrow side into the center of the circuit. In order to minimize reflections, the small dielectric plate can be provided with gradations at its ends.

Dielectrically loaded hollow waveguides are also used as polarization rotators or polarization transformers, with which the wave's polarization direction can be rotated, or, respectively, a plane polarization can be transformed into a circular polarization and *vice versa.*

The *rotating-plate phase shifter* is constructed similarly to the rotating supporting film attenuator pad (Fig. 3.22). In the middle of the rotatable $H_{11}$ circular waveguide cross section, a "$\lambda/2$ plate" of low-loss dielectric is located on the diameter plane. It is adjusted such that the partial wave undergoes a phase rotation with the electrical field parallel to it larger by $\pi$ than the partial wave with a perpendicular electrical field while passing through the circular waveguide section (because of the greater dielectric effectiveness of the plate in the parallel electrical field as a result of the lesser electrical discharge). This rotatable waveguide section of the rotating phase shifter causes, therefore, a change in the phase rotation at the angle $2\theta$ for a circularly polarized wave (in the case of a reversal of the rotation direction) with a rotation at angle $\theta$. At the middle circular waveguide section, transfers are connected to both sides at the input and output of the rectangular waveguide profile, in which an $H_{10}$-$H_{11}$- or an $H_{11}$-$H_{10}$-wave mode transformation takes place. The transformation of the incident plane polarized $H_{11}$-wave into a circularly polarized wave and the return transformation into a plane polarized wave at the input takes place with each of the fixed "$\lambda/4$ plates" located in the rectangular-circular transition at the input or the output, and which are in the vertical (longitudinal) direction on both sides of the rotatable $\lambda/2$-plate. If two waves in-phase and polarized perpendicular to each other propagate in a hollow waveguide, then we obtain a circularly polarized wave through the path delay of $T/4 = (\lambda_h/4)/v_p$; conversely, a circularly polarized wave can again be divided through such a path delay into two in-phase waves which are polarized perpendicular to each other. These $\lambda/4$ polarizations can be realized with the help of metal stems or dielectric inserts in the hollow waveguide. The $\lambda/4$ plates of the rotating phase shifter, which are also made of a low-loss dielectric material, are set at a 45° angle with respect to the $E$-field direction in the neighboring rectangular hollow waveguides, and produce the phase difference of $\pi/2$ necessary for a circular polarization between the partial waves with parallel and perpendicular electrical fields.

A trombone-shaped element is also used as a phase shifter, in which a U-shaped circuit component can be narrowly slid into two other circuits, and the effective circuit lengths and, therefore, the wave phase can be changed.

With another phase shifter, a rectangular hollow waveguide many wavelengths long can be divided along the center lines of the broad side. The hollow waveguide can be slightly bent over a screw fixture perpendicular to the slotted shaft. The propagation velocity and, therefore, the wave phase can be varied by changing the waveguide geometry created by this method.

Ferrite phase shifters will be described in Ch. 5.

## 3.7 PATH JUNCTIONS, TRANSITIONS (ADAPTERS), ROTATING JOINTS

Hollow waveguides of equal cross section are most often joined together with a *flanged coupling joint.* The flanges should show surfaces as smooth as possible and which lie on top of each other in order to pass the wall currents in the entire vicinity of the waveguide apertures that thrust against each other; otherwise, a mismatch is created and at a higher transported charge *spark-gap* breakdowns can occur. In order to compensate for insignificant discrepancies, a contact sheet made from a soft metal (most often beryllium-bronze, copper, or aluminum) can be laid between the flanges which border each other.

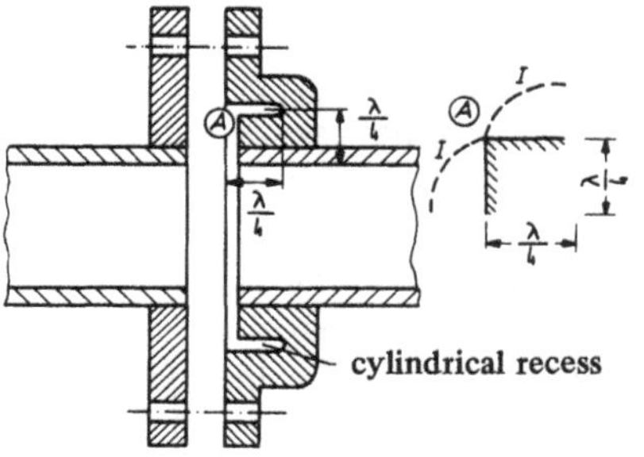


**Fig. 3.25** Choke flange connection

A reduction of high currents over the mechanical junction point of two hollow waveguides can be obtained with the choke flange connection. Figure 3.25 shows the principle of such a choke flange. With this we transform a short circuit in the waveguide inner wall at the connection point. Therefore, we provide the connecting flange with a narrow lateral pocket recess, whereby the end of the recess is located one-half wavelength from the inner surface of the waveguide. The $\lambda/2$ fissure creates the short-circuit transformation at its end to a short circuit at the inner surface of the waveguide. The mechanical contact point is located at one-half the slot length where a current node is present. This waveguide connection can only function at low surge in a narrow frequency range because of the slot length which depends upon the wavelength.

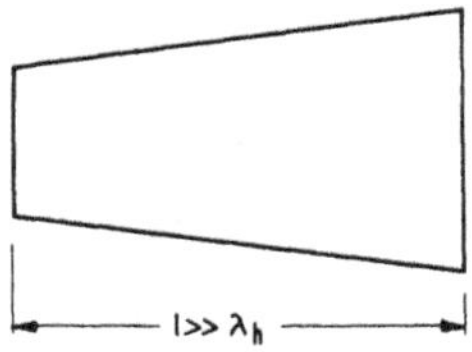


**Fig. 3.26** Linear transition (taper) between two different cross sections of the conducting wire

Connections between hollow waveguides with differing cross sections can be undertaken with a funnel-shaped transition (tapered; linear, according to Fig. 3.26; or exponential). In order to give this oscillation resistance transformation a wideband form, the length of the transition must be very large in comparison with the waveguide wavelength. A compensation of the contact points at the ends of the transition can be obtained by choosing its length to $l = n \cdot \lambda_h/2$, n is an integral. Such a transition can be undertaken to save more space with a $\lambda/4$ transformation (Fig. 3.18). The correct sizing of the transition is made more difficult by the field disturbances created at the transition points.

Transitions between hollow waveguides of differing cross sections (also between rectangular and circular hollow waveguides) can be performed by matching pads with layered dielectric inserts.

Two rectangular hollow waveguides to be connected can be rotated at a small angle with respect to each other in the case of a matching error which remains negligible, whereby, naturally, no apertures can be created. With a larger torque angle necessary for the rotation of the direction of polarization, usually 90°, we constantly use a rectangular hollow waveguide twisted on its axis, so-called *hollow waveguide rotations.*

In order to change the direction of hollow waveguides *line bends* and *line angles* are used. With angle plates, the corner is tapered in order to decrease the matching error. Usually, bends with a continuously running curvature and a constant cross section are used. The curvature, usually 90°, can be dealt with in the direction of the electrical lines of force (*E*-bends, *E*-angles) and, therefore, in the case of the rectangular hollow waveguide, over the broad side; or in the direction of the magnetic lines of force (*H*-bends, *H*-angles) and, therefore, with the rectangular hollow waveguide, in the direction of the narrow side.

We construct the rotation or curvature of a hollow waveguide over a length of $n \cdot \lambda_h/2$, $n = 1, 2, 3, \ldots$ (with a bend of medium arc length as a geometric center between the lengths of the inner and outer arcs). Therefore, the reflections at both ends of the transition compensate themselves and the mismatching through the transition remains negligible.

At the transition points between circuits with different wave modes, special *transitions (wave mode transformers)* are necessary. In order to excite a certain field mode, we need a coupling element in order to stimulate electrical fields (with a pin) or magnetic fields (through holes or slots), which run similarly to those of the field mode to be excited at the coupling point. The transition should guarantee as much as possible a reflection-free match and with as wide a band as possible.

The excitation of fields in waveguides often results from a coaxial transmission line. In this connection the hollow waveguide inner conductor can be lengthened so that it stimulates an electrical field. The inner conductor can also be mounted in the hollow waveguide in the form of a loop toward the outer conductor, and in this manner makes excitation across the electrical field possible (compare with couplings to resonators, sec. 3.12). In Fig. 3.27 some transitions between a coaxial transmission line and a hollow waveguide are outlined. Wideband matching can be attained by a corresponding sizing of the transition point. The transition between a coaxially directed wave and an $H_{11}$-wave in a circular hollow waveguide can be constructed in the same way as for an $H_{10}$-rectangular waveguide wave because of their similar field patterns.

In the case of a transition from a rectangular hollow waveguide to a microstrip line, according to M. Arditi, the hollow waveguide passes over gradually into a ridge hollow waveguide to which the strip line is soldered with the line strips on the ridge (equal breadths) and the datum plane on the waveguide's broad side.

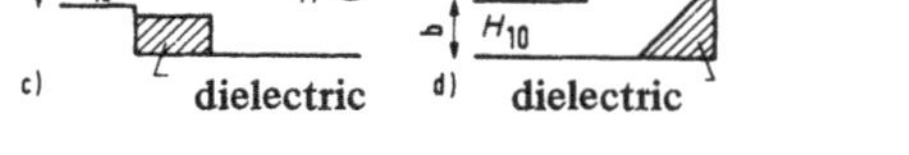

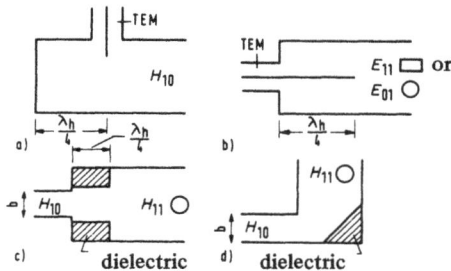

**Fig. 3.27** Transition between coaxial transmission lines and hollow waveguides, and rectangular and circular hollow waveguides

If, in a hollow waveguide, only the surface wave is excited and capable of propagating, then the higher field modes excited at the excitation point in order to meet the limit conditions are suppressed as attenuation modes from the coupling point outward. If, in the hollow waveguide to be excited, many wave modes are capable of propagation, then we can suppress undesired modes with *wave mode filters.* We can distinguish between absorbing and reflecting wave mode filters. *Absorbing* wave mode filters (or absorbers) represent active resistance to the undesired wave mode and, therefore, produce an absorbing effect. With this, we can strongly attenuate the undesired wave mode by mounting a dissipative conducting material (e.g., absorbing wires or supporting films) parallel to its electrical lines of force and perpendicular to those of the desired wave. *Reflecting* wave mode filters (or reflectors) form a reactive termination for the wave mode to be suppressed which causes reflection. We use conducting wires or plates in the direction of the electrical lines of force of the interferencing wave; or we hinder the propagation of the undesired wave mode with a suppression filter or through the transformation of a short circuit for those waves (e.g., with a resonance ring) at the excitation point. We also use dielectric plates which are laid out such that *E*-waves are allowed to pass reflection-free but *H*-waves are partially reflected. Through an arrangement of several such plates a mode filter can be realized.

As mode filters, we also designate arrangements used for the separate decoupling (or coupling-in) of various transmitted information channels (e.g., special directional couplers which exploit the differences in wave mode wavelengths) in a hollow waveguide with different independent wave modes.

By using the polarization directions of the same wave mode, which are perpendicular to each other in a hollow waveguide, we can double the number of transmission channels. Usually, we use the $H_{11}$-wave in the circular waveguide or the $H_{10}$-wave in the waveguide with a quadrangular cross section. For the separate coupling-in and decoupling, we use so-called *polarization filters.* Figure 3.28 shows a simple possibility for coupling two $H_{10}$-waves through rectangular hollow waveguides which are perpendicular to each other to a circular hollow waveguide, in which they are directed as $H_{11}$-waves with perpendicular polarization directions.

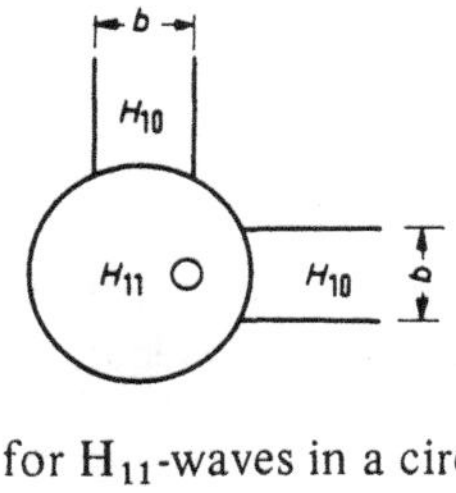


**Fig. 3.28** Polarization filter for $H_{11}$-waves in a circular hollow waveguide.

As a transition between fixed and rotating waveguides, we use *revolving turrets.* In order to maintain an output signal independent of the torque angle, we must use a line with a cylindrically symmetrical field distribution in the revolving turret, such as the coaxial transmission line or the circular hollow waveguide with an $E_{01}$-wave. In Fig. 3.29a, a revolving turret with rectangular waveguide connection ($H_{10}$-wave) is outlined as an example. The transition results from a coaxial transmission line section whose end in the rotating segment is formed as a dipole with a balance-unbalance transformer. It is important to note the capacitative transition at the interruption of the outer conductor (at A). In this case, the idling power at B is transformed over a distance of $\lambda/4$ into a short circuit at the conductor wall A, so that the transition of the longitudinal currents is guaranteed. The further $\lambda/4$-overlap connecting downward serves to shield the leakage field. Instead of the coaxial transmission line at the point of transition, a circular hollow waveguide with the $E_{01}$-wave can also be used. The excitation of the $H_{11}$-field mode of the circular hollow waveguide with its dynamically unbalanced field distribution must be suppressed, for example, through a $\lambda/2$ wave-trap or by exciting the $E_{01}$-wave over an apertured partition. The mismatch created by the use of the apertured partition can be eliminated by a slot diaphragm in the hollow waveguide ahead of the transition point.

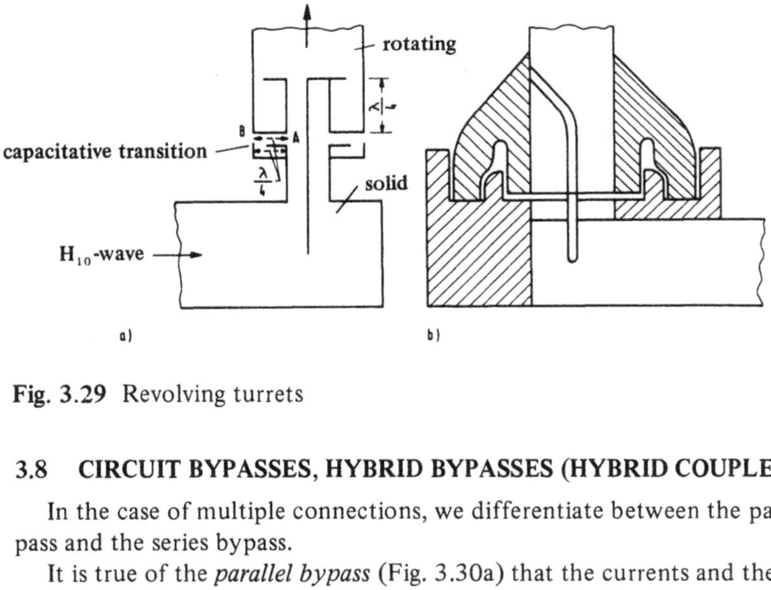


**Fig. 3.29** Revolving turrets

## 3.8 CIRCUIT BYPASSES, HYBRID BYPASSES (HYBRID COUPLERS)

In the case of multiple connections, we differentiate between the parallel bypass and the series bypass.

It is true of the *parallel bypass* (Fig. 3.30a) that the currents and the magnetic fields correspondingly distribute the loads at the branching point; whereas the voltage and the electrical fields are the same. For the main branch, the loads of the secondary branches transformed at the connection point appear to be connected in parallel.

In the case of the *series bypass* (Fig. 3.30b), the voltage or, respectively, the electrical fields of the main branch at the connection point are distributed in the secondary branches; whereas the currents and the magnetic fields are the same.

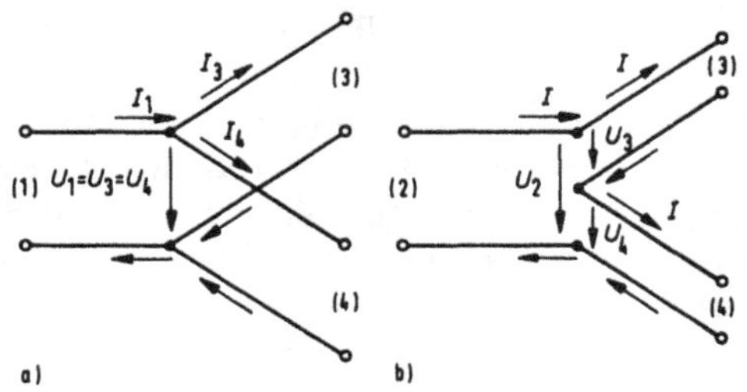


**Fig. 3.30** (a) Parallel bypass, (b) Series bypass

The loads of the secondary branches transformed at the connection point appear to be connected in a row for the main branch.

In the following, we will examine, as an example, bypasses of rectangular hollow waveguides, in which only the $H_{10}$-wave is capable of propagating for all branches. Circuit bypasses in stripline engineering will be found in Ch. 4.

A waveguide bypass in the plane of the magnetic lines of force – an *H-bypass* (Fig. 3.31a) – behaves similarly to a parallel bypass, and a bypass in the plane of the electrical lines of force – an *E*-bypass (Fig. 3.31b) – behaves like a series bypass. A charge which delivered to arm 1 of the *H*-bypass, or arm 4 of the *E*-bypass, distributes itself in equal parts to both other arms. With the *H*-bypass the signals in both output arms at an equal distance from the center of the bypass are in phase, and with the *E*-bypass out of phase. The input to the bypass is not adapted when both output arms are closed.

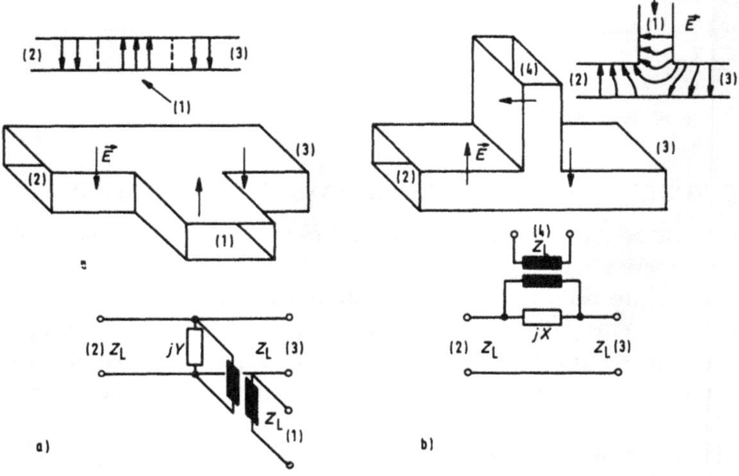


**Fig. 3.31** T-bypasses with simple equivalent circuit diagrams: (a) H-bypass, parallel-T-bypass, (b) E-bypass, series-T-bypass

In order to meet the limit conditions, higher order field modes occur. Since these are not capable of propagation in the connecting circuits, they only carry reactive power. This is expressed in the equivalent circuits by the dummy elements representing the field disturbance, or by $jX$. In order to compensate within a limited frequency range, and thereby improve the matching, diaphragms or pins can be introduced. In the bypass arm, we can attach short circuits such that no transmission or reflection-free transmission takes place between the other two arms (circuit breaker).

The combination of an $E$- and $H$-bypass leads to a quadruple-gate which is designated as an *E-H bypass* or a *hybrid-double-T bypass* (Fig. 3.32a). In the case of a symmetrical construction with the same termination at arms 2 and 3, a wave fed in arm 1 cannot excite a wave (of the surface wave mode) in arm 4 and *vice versa*; thus, no transmission takes place between gates 1 and 4; and, therefore, these gates are decoupled from each other ($S_{41} = S_{14} = 0$). As Fig. 3.32b shows, in the case of feeding an $H_{10}$-wave in arm 1, the electrical field is symmetrical to the median plane of the arrangement and, therefore, cannot excite an $H_{10}$-wave in arm 4. By feeding an $H_{10}$-wave in gate 4, the electrical field is out of phase, corresponding to Fig. 3.32c, at equal distances from the plane of symmetry in gate 1, and no $H_{10}$-wave is excited in arm 1. The bypass acts as a differential transformer (Fig. 3.32d). By feeding in gate 1 and a reflection-free termination of the remaining gates, signals appear in arms 2 and 3 with equal amplitude and phase, and in arm 4 no signal is coupled in. A wave entering arm 4 excites waves of equal amplitude and opposite phase, whereas in gate 1 no signal appears.

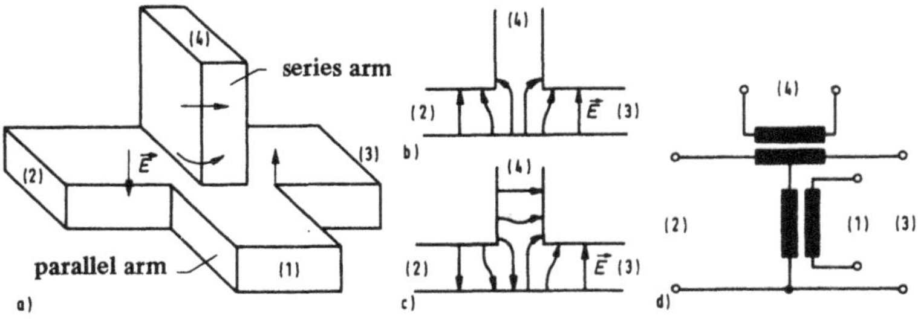


**Fig. 3.32** (a) hybrid-T-bypass, combined parallel and series bypass, E-H-bypass, (b) E-field at feeding in gate 1 and (c) in gate 4, (d) differential transformer as an equivalent circuit

A quadruple-gate circuit, in which, in the ideal case, a charge fed into a gate distributes itself to two additional gates and no charge is coupled in the fourth gate is also known as a *hybrid bypass.*

In the case of reflection-free termination at arms 2 and 3, we obtain mismatch for gates 1 and 4. These can be eliminated in arms 1 and 4 with matching elements (by installation of metal diaphragms and pins), so that (approximately) $S_{11} = S_{44} = 0$. For a non-dissipating bypass, $S_{22} = S_{33} = 0$ and $S_{23} = 0$ because of unity (see Ch. 2). The quadruple-gate can, therefore, be operated adaptively to all four gates, and gates 2 and 3 are decoupled from each other as well. The bypass is laid out such that matching dominates with an adaptive termination of three branches at the fourth gate. A wave entering gate 2 is then distributed in equal parts to arms 1 and 4, just like a wave fed into arm 3. This is $|\underline{S}_{12}| = |\underline{S}_{24}| = 1/\sqrt{2}$. Thus, we obtain a directional coupler (sec. 3.10). If the charge fed in a gate distributes itself in equal parts to two other adaptive gates, as with this bypass (coupling attenuation 3 dB), then we speak of a 3-dB directional coupler. A bypass to be operated adaptively to all four gates like the circuit described, is called a *decoupled bypass,* and specifically the illustrated double-T-hybrid circuit *magic T.*

If we choose the planes of reference in arms 1 and 4 such that $S_{12}$ and $S_{24}$ are real, then we obtain for the S-matrix of an ideal magic T

$$S = \frac{1}{2}\sqrt{2}\begin{vmatrix} 0 & 1 & 1 & 0 \\ 1 & 0 & 0 & 1 \\ 1 & 0 & 0 & -1 \\ 0 & 1 & -1 & 0 \end{vmatrix} \tag{3.26}$$

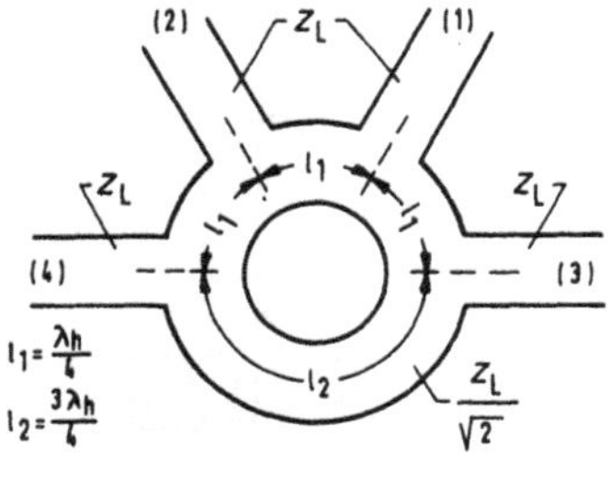


**Fig. 3.33** Hybrid ring bypass

(3.26)

In Fig. 3.33, the diagram of a *decoupled ring bypass* or *hybrid ring bypass* is represented. By feeding in gate 1 of the hybrid ring, the signal is distributed in equal parts to gates 2 and 3 with equal phase, and by feeding in arm 4 with a phase shift of 180°. If we feed in gate 2, and gates 1 and 4 are reflection-free (or mismatched in the same manner), then at gate 3 no signal appears since the waves circulating in both directions have paths differing by $\lambda_L/2$ up to gate 3, and cancel each other out because of the phase opposition. Gates 2 and 3 are, therefore, decoupled from one another; the same holds true for gates 1 and 4. The arrangement can, therefore, be represented in an equivalent-circuit diagram by a differential transformer. For the S-matrix of the ideal structural component, we obtain with a corresponding distance of the gates from the ring (S-parameter is real) the same S-matrix, Eq. (3.26), as for the magic T. In order to achieve matching, the oscillation resistance of the arms of the waveguide-ring hybrid, which are connected in series to the ring (with the stripline-ring hybrid, parallel, Fig. 4.10c), must amount to $\sqrt{2}$ times the oscillation resistance of the ring.

$$Z_A = \sqrt{2} \cdot Z_R \ . \tag{3.27}$$

Since the distances of the ring hybrid arms are fixed with respect to wavelength, the ring conduit does not have as wide a band as the magic T.

Hybrid bypasses are capable of many possible applications. They can be used in the construction of *symmetrical mixers* (balance mixers). Figure 3.34 shows the principle of such a push-pull mixer. Arms 2 and 3 of the hybrid are closed with low reflection by mixer diodes. The input signal is fed in gate 4 and the superimposed mixing oscillator signal is fed in gate 1. The input circuit is, therefore, thoroughly decoupled from the oscillation circuit (no emission of the oscillator signal over the receiver input). The input signal is distributed to both mixer diodes with opposite phase, whereas the mixer oscillator keeps the diodes on course and in phase. Therefore, the ZF-signals created are of opposite phase and the noise voltage components of the mixer oscillator are in phase at both diodes. If at this point both out of phase ZF-signals are brought together in a push-pull transformer, then the signal voltages are added together whereas the noise of the superposition oscillator cancels out.

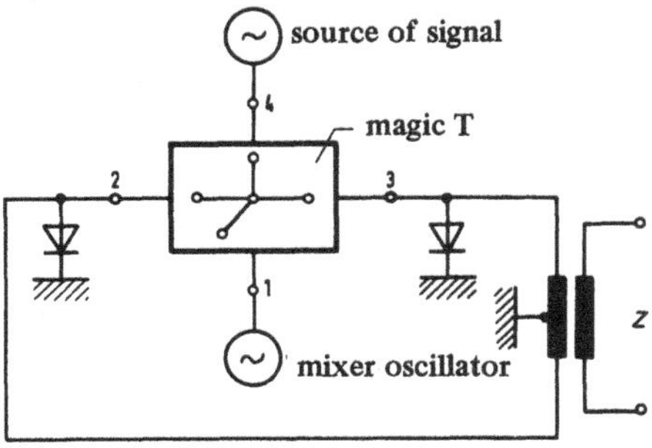


**Fig. 3.34** Push-pull mixer

If gates 1 and 4 of the magic T are closed with adjustable short circuits (tie lines), then we obtain the *E-H* tuner which is used as a matching element. By adjusting the lengths $l_1$ and $l_2$ of the short-circuited lateral arm, the corresponding parallel or series reactances in the circuit are transformed and the impedance $\underline{Z}_2$ can be adapted to the waveguide oscillation resistance $Z_L$.

The *E-H* bypass can also be used similarly to a measuring bridge in order to *measure the impedance.* In this connection, gate 1 is fed into, and at gates 2 and 3 a variable calibrated impedance and the impedance to be measured are connected. At gate 4 a low-reflection indicator is attached, which yields a zero reading in the case where the impedances are equal (amount and phase).

With a hybrid bypass, a *phase shifter* may also be constructed. By feeding the hybrids (Fig. 3.32a) in arm 1, the waves appearing in arms 2 and 3 are in phase at the same distance from the plane of symmetry for this arrangement, and by feeding in arm 4 they are out of phase. In order to operate it as a phase shifter, we can mount, for example, a shorting plunger at a distance $l$ or $l + \lambda_L/4$ from the plane of symmetry in branches 2 and 3. The wave fed in arm 1 is reflected at both short circuits. The reflected waves have a phase shift of 180° in the plane of symmetry because of the shorter path by $\lambda_L/3$ in arm 3. Therefore, they cancel each other out in branch 1, whereas in arm 4 they are added in phase. If the length $l$ is subsequently increased by $\Delta l$ in branches 2 and 3, which means the path length difference of both partial waves is increased by $2\Delta l$, then at gate 4 we obtain a phase shift of

$$\Delta\varphi = \frac{2\pi}{\lambda_L} \cdot 2\,\Delta l\,. \tag{3.28}$$

Additional applications of hybrid bypasses are obtained by use as circuit dividers or in transmitter-receiver switches (duplexer; e.g., in radar devices). They are also used in the formation of the sum and difference of high-frequency signals (e.g., with monopulse radar): if two signals of the same frequency with amplitudes $A_1$ and $A_4$ are fed in gates 1 and 4, then at gate 2 we obtain the sum signal $A_2 = 1/\sqrt{2}(A_1 + A_4)$ and at gate 3 the difference signal $A_3 = 1/\sqrt{2}\,(A_1 - A_4)$ (or *vice versa* depending on the phase of the input signal).

In the case of feeding into gates 2 and 3, sums or differences appear in gate 1 or gate 4, respectively.

## 3.9 COUPLED HOLLOW WAVEGUIDES

With a coupling aperture in a waveguide wall, a portion of the energy can be transfered to a second neighboring hollow waveguide. To this end, we usually use hole coupling with a small round coupling hole, or slot coupling with a narrow rectangular window.

In Fig. 3.34, the path of the electrical lines of force with electrical coupling and the path of the magnetic lines of force with magnetic coupling are represented for *hole coupling* through a hole in a metal wall. In the case of electrical coupling, the electrical lines of force come in perpendicular to the coupling hole on the secondary line; the electrical coupling can be represented in an equivalent circuit by an electrical dipole. With magnetic coupling, the magnetic lines of force curve through the hole into the adjoining area; the magnetic dipole is its equivalent replacement representation.

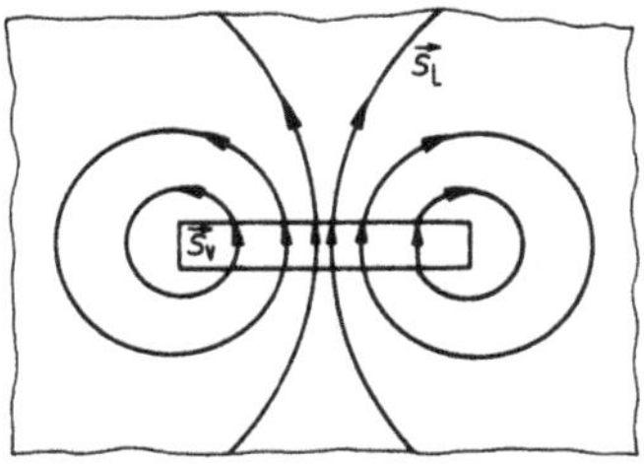

**Fig. 3.35** Displacement current lines $S_V$ and wall current lines ($S_L$) on a resonance slot

We often also use slot coupling with a narrow rectangular slot. The behavior of such a window is similar to that of the resonance diaphragms dealt with in sec. 3.1. Figure 3.35 shows the shift current distribution (electrical lines of force) and the wall currents on a resonant slot. The slot is mounted on the waveguide wall such that the broad sides of the slot interrupt the path of the wall currents. Then a portion of the current transversely directed to the broad side flows around the slot. The other portion flows as a shift current straight through the slot; with this an electrical field is connected at right angles to the slot and a magnetic field running perpendicular to it is interconnected. For an effective stimulation, we choose a slot length $l \approx \lambda_0/2$ at which resonance occurs. A portion of the energy is radiated in the secondary line over the electromagnetic field in the slot. The intensity of the decoupled energy can be influenced by the effective length of the slot with reference to the wall current losses. Since the slot is excited by the currents running diagonally to it, the coupling increases with the size of the wall current portion, which is directed normally to the slot's broad side. If the slot cuts across the current paths (cross slot) with its broad side in the propagation direction of the hollow waveguide, then we obtain a series resistance coupled in from the secondary line to the main shaft. If the slot is excited by wall currents running transverse to the propagation direction (longitudinal slot), then a parallel conductance is coupled into the main line. We speak of *series* or *parallel coupling* of the hollow waveguide depending on whether they interrupt longitudinal or cross currents of the slot. The impedance coupled in is formed in the case of a secondary line with a cross slot (longitudinal slot) from the series connections (or parallel connections) of the terminal impedances transformed at the coupling location on both sides of the secondary line. At the resonant frequency, the secondary or series impedances coupled into the main line are real. In Fig. 3.36 two examples of slot couplings are represented: (a) series coupling, where in both hollow waveguides the longitudinal currents are interrupted, and (b) a parallel series coupling, in which the cross currents of the main line are coupled to longitudinal currents of the secondary line.

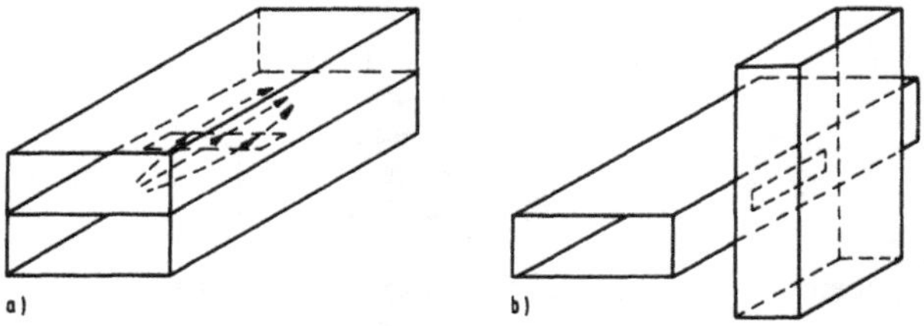

**Fig. 3.36** (a) series-series coupling, (b) parallel-series coupling

In the secondary line coupled to the main line, waves propagate from the coupling point outward on both sides. If the energy only propagates in one direction in the secondary line as well, then we use a directional coupler.

## 3.10 DIRECTIONAL COUPLERS

Directional couplers (circuit couplers) are structural components in which waves from a line (main line) are coupled into a second, usually parallel, line (secondary line). The coupling may take place at one or two distinct transfer substations, for example, through coupling holes in waveguides lying on top of each other, through a series of such coupling apertures, or continuously (distributed coupling) along a coupling distance. From a wave propagating in the forward (incident) direction in the main line, parts can be coupled in the secondary line which propagate in the backward (reflected) direction (backward coupling-in), or the forward direction (forward coupling-in), or to both sides (non-directional coupling-in) (see Fig. 3.37b).

Figure 3.37 shows the diagram of a directional coupler with the chosen designations of its terminal pairs or gates. An ideal directional coupler is the quadruple-gate (or four-port junction) for which the active power directed at a gate is only distributed to the receiver at two gates, whereas no power is lost at the fourth gate. According to Fig. 3.37a, the active power $P_1$ of the wave fed into gate 1 of the ideal directional coupler divides itself into the active powers $P_2$ and $P_3$, and is distributed to the receiver adapted to the circuit oscillation resistances

$Z_{L_1}$ or $Z_{L_2}$ at gates 2 and 3; no power is transmitted to gate 4. By feeding in gate 4, a transmission also only takes place to gates 2 and 3. Gates 1 and 4 are decoupled from each other. If gate 2 or gate 3 is fed into, then there is only transmission to gates 1 and 4; hence, gates 2 and 3 are decoupled. When three gates of an ideal directional coupler become adapted closed, then it is also adapted at the fourth gate.

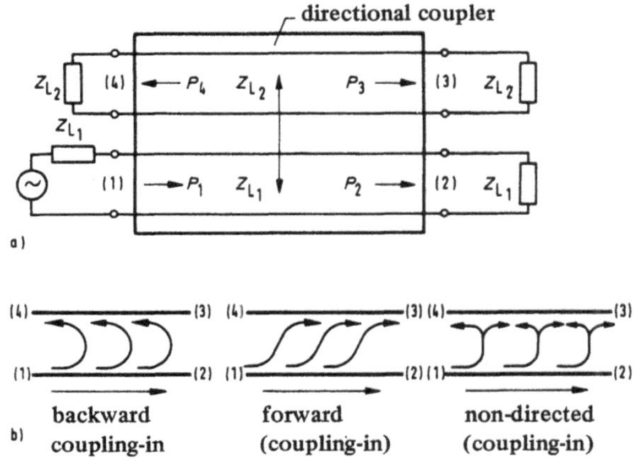


**Fig. 3.37** (a) diagram of a directional coupler, (b) coupling forms

In the S-matrix of the ideal directional coupler $S_{11} = S_{22} = S_{14} = S_{23} = 0$; the S-matrix of the non-dissipative directional coupler is unitary with $S_{33} = S_{44} = 0$, $|\underline{S}_{13}| = |\underline{S}_{24}| = |\underline{S}_{12}| = |\underline{S}_{34}|$ and $|\underline{S}_{12}|^2 + |\underline{S}_{13}|^2 = 1$, $|\underline{S}_{12}|^2 + |\underline{S}_{24}|^2 = 1$. Through a suitable choice of the plane of reference in arm 1, $\underline{S}_{12}$ is a positive real magnitude $A$, and through a proper choice of the plane of reference in arm 4, $\underline{S}_{34}$ is real; as a result of the suitable choice of the plane of reference in arm 3, $\underline{S}_{13}$ has a positive imaginary value j$B$. Then the following is valid for the S-matrix of the ideal directional coupler:

$$\underline{S} = \begin{vmatrix} 0 & A & \mathrm{j}B & 0 \\ A & 0 & 0 & \mathrm{j}B \\ \mathrm{j}B & 0 & 0 & A \\ 0 & \mathrm{j}B & A & 0 \end{vmatrix} \tag{3.29}$$

with $A^2 + B^2 = 1$.

The zero-elements in the principal diagonal specify that all gates are adapted. The additional zero-elements indicate that gates 1 and 4, as well as 2 and 3, are decoupled from each other.

The portion of the desired power coupled into the secondary line ($P_3$ in gate 3) in relation to the power fed into the main line ($P_1$ in gate 1) in the case of a reflection-free closed main line is designated as the *coupling attenuation*

$$\frac{a_K}{dB} = 10 \log \frac{P_1}{P_3} \ . \tag{3.30}$$

The specification of the usually negligible power ($P_4$) appearing at a reflection-free shut-off of the main line, with a real directional coupler also at the decoupled gate of the secondary line (gate 4) referred to the power ($P_3$) coupled in the desired direction, serves the *directional attenuation*

$$\frac{a_R}{dB} = 10 \log \frac{P_3}{P_4} \ . \tag{3.31}$$

The 3-dB coupler is customary with $P_3/P_4 = 0.5$, as well as the 6-dB coupler with $P_3/P_1 = 0.25$, the 10-dB coupler with $P_3/P_1 = 0.1$, and the 20-dB coupler with $P_3/P_1 = 0.01$. The coupling attenuation should be as high as possible, usually in the range of approximately 20 to 50 dB.

A primary application of directional couplers is the separate measurement of incident and reflected line waves (reflectometer), and thereby the determination of S-parameters (reflection and transmission factors) and apparent resistances. For the power transported by the incident wave (from gate 1 to gate 2) in the main line, we obtain a proportional power at one gate of the secondary line (gate 3), and at the other gate (gate 4) for the reflected wave. Gates 3 and 4 of the secondary line are to be closed reflection-free so that the measured values at each other gate are not tampered with by the reflections created there. For more specific measurements, we often use the separate three-armed directional coupler for measurement of the incident and reflected power. The directional attenuation of a directional coupler is an indication of the precision with which small reflectivities (of gate 2) can be measured.

Directional couplers also serve the reflection-free decoupling of signals as attenuator pads or phase shifters. The 3-dB coupler is also used as a power divider in the distribution of power to equal parts, or in the combining of the power from two sources; for example, in the case of amplifiers. An additional possible application of directional couplers is as a matching element. It is also used in antenna change-over switches.

In the following subsection, a few type models of directional couplers in waveguide technology are introduced. Directional couplers in stripline technology are to be found in Ch. 4.

### 3.10.1 Dual and Multiple-Aperture Couplers

The mode of operation of the dual-aperture coupler is the easiest to understand. Figure 3.38a shows a dual-aperture coupler in waveguide execution. The coupling of the secondary line 3-4 to the parallel main line 1-2 takes place through two small coupling apertures separated by $\lambda_h/4$ (slots or round holes) in the rectangular waveguide broad sides bordering each other. At both coupling apertures a portion of the propagating wave energy in the main line is coupled into the secondary line. Waves are thereby excited in the secondary line which move out from both transfer substations toward either side. The principle of electrical and magnetic coupling in the case of hole coupling through an aperture in a metal wall is illustrated by Fig. 3.38b with the path of the electrical or magnetic lines of force. In the case of a tight coupling, the wave undergoes a phase shift of 90°. With a wave moving in the forward direction (from gate 1 to gate 2) in the main line, the partial waves coupled in the secondary line which move in the forward direction (to gate 3) and that proceed from the coupling apertures, add since they are of equal phase as a result of their virtually equal path length. The partial waves moving in the backward direction (to gate 4) are out of phase because of their path difference of $2\lambda_L/4 = \lambda_L/2$ and cancel (compare with Fig. 3.38b, tightly coupled field portion corresponding to coupling factor $k$, at small $k$ with $1 - k \approx 1$ amplitude equality of the partial waves to be superimposed). Because the waves propagate in the same direction in the main and secondary lines such directional couplers are designated as *forward-wave couplers.*

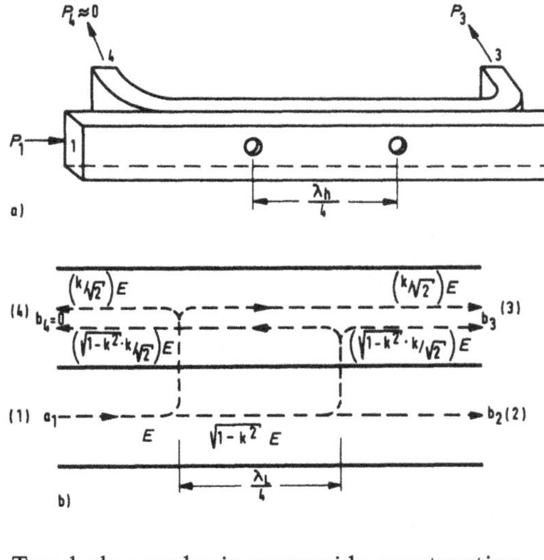


**Fig. 3.38** Two-hole coupler in waveguide construction

In another use of directional couplers, two rectangular hollow waveguides are arranged crossing each other by 90° and are coupled over two common round apertures located at a distance of $\lambda_h/4 \cdot \sqrt{2}$ from each other in the broad sides, whereby the connecting line of the apertures in each case is below 45° with respect to the waveguide directions. The coupling apertures of such a *cross coupler* can also be designed as two cross slots in the waveguide broad sides (Fig. 3.39). The partial wave coupled in the secondary line 3-4 and propagating toward gate 3 are added in phase and the partial waves propagating toward gate 4 compensate because of the $\lambda_h/2$ path difference.

Because of the distance $\lambda_h/4$ of the coupling holes of the dual-aperture coupler, its operating bandwidth is not very high. Higher bandwidths can be attained with a larger amount of coupling holes and slots, with *multiple-aperture couplers.* In the case of a distribution of the individual coupling intensity (through different aperture sizes or distances) according to the binomial coefficients (e.g., 1:4:6:4:1), we obtain a maximally flat characteristic in the operating frequency range. If we grade the coupling factor according to a Chebyschev distribution, then we obtain a characteristic with a constant ripple.

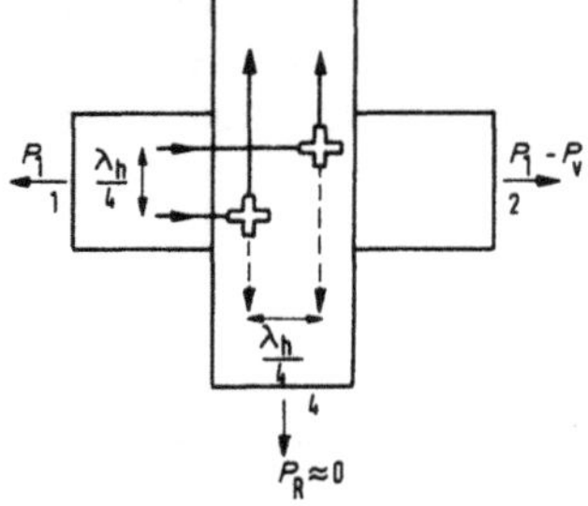


**Fig. 3.39** Cross hole coupler

### 3.10.2. Directional Coupler with Distributed Coupling

We obtain a so-called distributed coupling if we pass over from the distinct transfer substations of the multiple-aperture coupler to a continuous coupling along a coupling distance $l$ in the direction of propagation. Thus, we also obtain directional couplers with good wideband characteristics. The coupling distributed over a distance can take place, for example, in the arrangement of a vertical slot in the common broad side waveguide wall (vertical-slot coupler) according to Fig. 3.38a. If, with this, the coupling length amounts to $l = \lambda/2$, then the effect of the arrangement, which also works as a forward-wave coupler with the dual-aperture coupler mode of operation described in sec. 3.7.1, can be explained by the interaction in each case of two coupling amounts at a distance of $\lambda/4$.

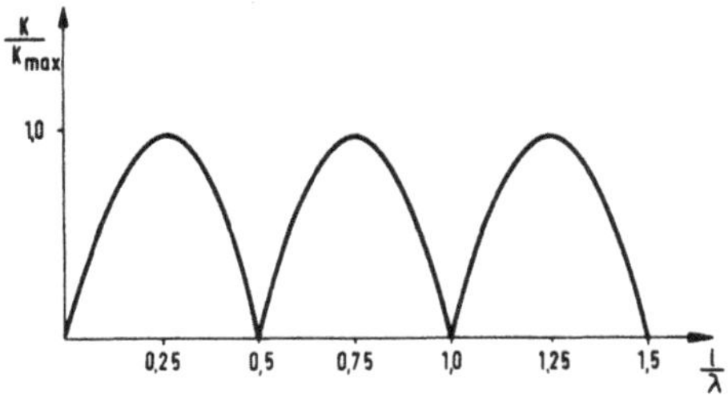


**Fig. 3.40** Normalized coupling transmission factor as a function of the coupling distance l relative to the wavelength

In Fig. 4.9 the stripline design of a directional coupler with distributed coupling is represented. It operates with a coupling length of

$$l = (2n + 1) \cdot \frac{\lambda}{4}, \quad n = 0, (1, 2, \ldots) \tag{3.32}$$

as a backward-wave coupler. Through a coupling dependent on the location, the bandwidth of the directional coupler can be essentially increased; for example, through an exponential increase of the conductor distance and width.

Directional couplers with distributed coupling between parallel lines were first described in 1941 by Pistolkors and Neumann.

In Fig. 3.40 the path of the normalized coupling transmission factor $k/k_{max}$ is illustrated graphically with $k = A_1/A_3 = \sqrt{(P_1/P_3)}$, e.g., $k_{max} = \sqrt{2}$ in the case of the 3-dB directional coupler for constant coupling.

An integration of the coupling parts along the coupling distance, for the case of a constant coupling according to amount and phase within the coupling interval between the main and secondary line for the directional effect, yields

$$\frac{P_3}{P_4} = \left(\frac{2\pi l/\lambda_L}{\sin 2\pi l/\lambda_L}\right)^2 \tag{3.33}$$

with $l$ = coupling length, $\lambda_L$ = line wavelength.

We obtain pole locations for $P_3/P_4$ at $l/\lambda_L = 0.5, 1, 1.5, \ldots$ with minima lying in between, and with increasing $l/\lambda_L$, growing minima. We obtain, for example, with $l = 2\lambda_L$ a wideband directional coupler with $10 \log P_3/P_4 \geqslant 21$ dB (corresponding to the minimum lying between $l/\lambda = 1.5$ and 2.0).

In the case of a coupling triangular path, which is zero at the range limits, we obtain

$$\frac{P_3}{P_4} = \left(\frac{\pi\, l/\lambda_L}{\sin \pi\, l/\lambda_L}\right)^4 . \tag{3.34}$$

We now obtain for $l = 2\lambda_L$ a wideband directional coupler with $10 \log P_3/P_4 \geqslant 35$ dB.

The effect of the distributed coupling can be approximated by closely neighboring individual coupling apertures.

### 3.10.3 Single-Aperture Couplers

Occasionally, the single-aperture coupler according to Bethe is also used. With this directional coupler, two rectangular hollow waveguides with $H_{10}$-wave propagation are coupled together over a round aperture in the middle of the common broad side. A wave is excited in the secondary line through the coupling with the electrical field as well as through the magnetic coupling (see also hole coupling, Fig. 3.34d). Both partial waves should cancel out in one direction (direction of the main wave in the main line) and intensify in the other direction. In order to do so, both partial waves must be equally strong. Since the magnetic coupling dominates through the hole, we rotate the waveguide axes at an angle of 60° with respect to each other and do not arrange the coupling aperture in the middle of the waveguide broad sides. With the rotation, the electrical coupling takes place, which can be equivalently represented by an electric dipole independent of angle $\theta$, and the magnetic coupling, which can be described by an equivalent magnetic dipole, takes place proportional to $\cos \theta$. In this manner, the magnetic coupling is decreased corresponding to the directional effect of an equivalent magnetic dipole, whereas the electrical coupling remains uninfluenced because of the omni-directional characteristic in this plane.

A disadvantage is the relatively low directional attenuation of the Bethe directional coupler.

## 3.11 MICROWAVE INTERRUPTERS

Microwave interrupters can be constructed for either electronic or mechanical operation. In electronically operated interrupters, we often use PIN-diodes (see *Mikrowellentechnik*, Vol. 2, Ch. 3) whose HF resistance independent of the flowing direct current is directed for either input or output conditions. The diode can, for example, in a closed condition, allow activation of a resonance pin in the waveguide, at which the approaching wave is reflected. In the case of an interrupter constructed with a magic T, arms 2 and 3 (Fig. 3.32a) are closed with semiconductor diodes. With the help of its initial stress, symmetrical arrangement can be attained and thereby an obstruction between arms 1 and 4; or, in the case of asymmetry, a through-connecting extension. We also often exploit the attenuation characteristics of microwave ferrites which change with the premagnetization (see ferrite attenuator pads, Ch. 5) in the construction of electronic microwave interrupters.

A mechanically operated microwave interrupter can be formed by a curved or angular waveguide section which is rotated between the hollow waveguides to be connected, or is mounted between the approaching hollow waveguide and an adapted load. The electrical contact (for the wall currents) is produced by the transformation of short circuits in the waveguide separations to be connected (short circuited $\lambda/2$ line). In another possibility, an $E$-plane T-connection is used, whereby the energy directed into an arm is reflected from the undesired arm by a resonance ring placed crosswise and guided into the other arm whose ring is adjusted so that it causes matching.

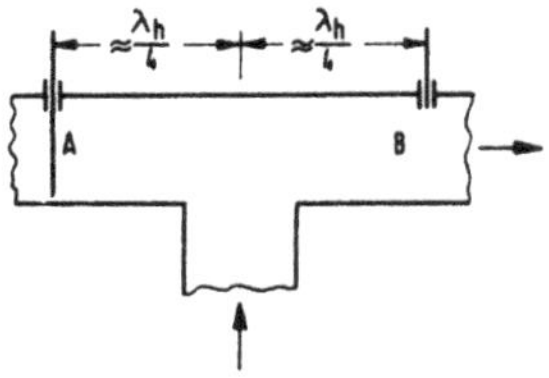


**Fig. 3.41** Microwave change-over switch

In Fig. 3.41 the principle of a change-over switch is represented which does not need an intermediate part. The incident wave is guided, respectively, by a short circuit or an idling power in one branch (at A or B) as reflection-free as possible onto the other branch line. The short-circuit plane is situated at a distance of approximately $\lambda_h/4$ from the center of the bypass. The short circuits can be realized mechanically with walls, or electrically with variable resistances, for example, with diodes. The short circuit or idling power can also be transformed over an additional lateral hollow waveguide in the corresponding branching arm.

# CHAPTER 4

# STRIPLINE COMPONENTS AND MICROWAVE INTEGRATED CIRCUITS

## 4.1 THE TECHNOLOGY OF MICROWAVE INTEGRATED CIRCUITS (MIC)

We designate striplines (see also Ch. 1) as those high-frequency lines in the form of flat conducting strips parallel to *a* (in the case of an asymmetrical circuit), or (in the case of a symmetrical circuit) between two long thin conducting planes and with a dielectric column support (substrate). Because of the relatively high attenuation, striplines are not suitable for longer lines; the charges that can be transmitted are also negligible. They are introduced as connectors in the form of stripline components in microwave integrated circuits. Because of their simple construction, the asymmetrical circuit design or the microstrip line is favored. Similarly, as in coaxial or waveguide technology, many microwave circuit components can be constructed with striplines. They have the advantage of negligible occupied space and small weight. Stripline technology is, therefore, suitable for the construction of microwave integrated circuits and to mass production. Microwave integrated circuits are abbreviated as MIC (or plural MICs). Through the integration of many components on the carrier plate, we obtain the advantages of mitigating the need for long junction lines, a smaller number of connectors and, therefore, of contact points. Besides this, very wideband circuits can be realized because of the negligible dispersion (this means that phase velocity is slightly frequency dependent) in the surface wave operation. With stripline circuits, we obtain high reliability and good reproduction. Disadvantages are the higher losses in comparison with coaxial or waveguide technology (lower efficiency of dummy elements), and smaller transferable charges.

The development of MICs was introduced in 1965 in the area of phased-array antennas. Examples of MICs are mixers, amplifiers, oscillators, receiver circuits, Doppler radar circuits, PIN-diode interrupters and phase shifters, ferrite phase shifters, and phase modulators.

Microwave integrated circuits can be manufactured by thin-film or thick-film technology. In microwave technology the two techniques are not differentiated from the manufacturing method of low-frequency circuits using the thickness of the conducting layer, which here amounts to (at least) 3 to 5 times the penetration depth.

In *thin-film technology,* we mount the conducting layer in the vacuum on the column support through evaporation or cathodic disintegration. As conductor material, we use copper, silver, gold, or aluminum. First, a very thin, highly resistive metal adhesive layer (thickness $\ll$ penetration depth) made of nickel-chrome, chrome, titanium, or tantalum for improved adhesion is mounted on the substrate (adhesion through a layer of oxide between the substrate and the conductor). The conductor layer, which is evaporated or dusted on it, is finally galvanized to the necessary thickness (approximately 3 to 5 depths of penetration). The line model can be manufactured by way of photolithography and etching.

In *thick-film technology,* the conducting layer is applied as a metallic powdered paste to the substrate and the film is then burned. The conductor structures can be manufactured with the help of photo-etching technology, or it can be printed with the conductor paste by silk-screen technology through a finely meshed web with the help of masks into which the line model is worked. Usually, thick-film microwave circuits exhibit high losses because of negligible electrical conductance of the burned conductor layer, porosity of the film, and insufficiently clean edges of the printed conductor paths. This technology is suitable for cost-effective mass production and is used up to approximately 15 GHz.

Microwave integrated circuits can be constructed in hybrid or monolithic form. In a *hybrid* integrated circuit, the semiconductor components (diodes and transistors) and also eventually the resistors and concentrated capacitors and inductors, are not integrated with the substrate supporting the conductor structure, but rather are subsequently added (hence the term *hybrid* or mixed). The semiconductor elements of the hybrid circuit can be realized with stripline terminals, low-power diodes in beam-lead technology with thin terminal bands (see *Mikrowellentechnik,* Vol. 2, Ch. 3), and at higher frequencies (millimeter waves) we use "naked" semiconductor chips, which can be attached with bonding wires or bands. Hybrid circuit resistors, capacitors and inductors can be introduced, as well as diodes and transistors, as finished components (hybrid elements) in the circuit, and like these can be attached by soldering, thermo-compression bonding, or with the help of conducting adhesives. Passive components can, however, also be directly produced (integrated) by the stripline material, eventually through the application of additional insulating layers (e.g., $SiO_2$). Hybrid circuits are constructed in thin-film or thick-film technology.

With *monolithic* integrated circuits, the entire circuit is contained within a single semiconductor component; the semiconductor elements are also realized in the semiconducting substrate of the connecting stripline. As semiconductor material, silicon (e.g., circuits with impatts) and, usually gallium arsenide are used, since the latter is well suited to microwave semiconductor elements (e.g., GaAs MESFET, GaAs Schottky-diodes, and variable reactor diodes [see *Mikrowellentechnik,* Vol. 2, Chs. 3 and 4]) and the insulating semiconducting

GaAs layer is very highly resistive. On the highly resistive semiconductor substrate, we produce the necessary semiconducting layers with an epitaxial growth or implantation. Thin-film resistors, condensors, and inductors are formed on the same substrate as the semiconductor elements. Stripline structures can be realized with an insulating GaAs layer as dielectric between the conductor paths and a highly doped, well conducting GaAs semiconductor layer as datum plane, which is bonded on the outside for the contact. The metal linings for conducting connections are produced as in the case of the thin-film hybrid integrated circuits. In manufacturing the circuits, we use photolithography techniques and introduce ion implantation in order to produce precise, narrowly confined dopings.

Monolithic microwave integrated circuits can be constructed upon very small substrate surfaces with concentrated elements. Besides, with concentrated elements, a wider range of impedance values can be more easily produced than with line connecting elements. In order to increase the realizable range of values for passive components, partially active elements can be introduced. Another advantage of monolithic microwave circuits is that many fewer bonding wires are necessary for connections; the wiring arrangement of semiconductor elements directly in their vicinity brings with it – especially in the case of millimeter waves – the advantage of fewer losses and parasitics. Therefore, in the case of wideband amplifiers with matching elements close to the active elements, we obtain higher bandwidths. Monolithic circuits possess good reproducibility and are very well suited to mass production. Disadvantages are the relatively low efficiency of the dummy elements and the fact that the circuits cannot be subsequently trimmed. Examples of monolithic circuits are amplifiers, oscillators, mixers, phase shifters, modulators, and interrupters. Monolithic circuits have already been manufactured for frequencies around 100 GHz.

Since integrated microwave circuits are difficult to trim once manufactured, a more precise design with the help of a computer (so-called computer aided design or CAD) is necessary.

Capacitors and inductors of such microwave circuits can be carried out as so-called *distributed elements* (line connecting elements) or as *concentrated* or, respectively, quasi-concentrated *elements* (lumped elements).

In the case of distributed operating circuit structures, dummy elements (capacitors, inductors, resonant circuits) are realized with a length comparable to the wavelength and are determined by the (idling or short-circuited) portions of the line element (see Ch. 1). The desired input reactance comes about through the fields divided along the line.

Coils and condensers can, however, also be produced like the resistors as concentrated or lumped active elements. We say that a circuit component is concentrated if its dimensions remain small as compared to the wavelength used and, therefore, the phase along the component is almost constant. The values of such components remain independent of frequency in the frequency range of interest. Concentrated components such as resistors, capacitors, and inductors can be integrated in the circuit or laid out for subsequent installation as chip components, which are soldered, bonded, or glued in the circuit. Concentrated circuit components which are to be subsequently introduced into the circuit are also called *hybrid elements* like the semiconductor elements used. Concentrated integrated components are produced in thin- or thick-film technology. With the help of photolithography or thin-film technology, the dimensions of resistors, condensers, and inductors can be kept small enough so that such elements can be operated up to approximately 50 GHz. Concentrated components offer the advantage that they can be smaller than distributed components and that with them smaller circuit dimensions can be obtained; for large quantities this is also cheaper since more circuits can undergo the same production steps.

Ferrite substrate circuits can be constructed with gyromagnetic components (Ch. 5) such as circulators, directional circuits, reciprocal and non-reciprocal phase shifters, attenuators, interrupters, modulators, limiters, and tunable filters. If a circularly polarized high-frequency magnetic field is necessary, then the *slot line* and the *coplanar line (double-slot line)* (Ch. 1) are particularly well suited. The surface wave of the slot and coplanar lines is in the first approximation of the *H*-mode. Thus, the magnetic field shows cross and longitudinal components and, thereby, we obtain areas in which the magnetic *HF*-field is circularly or elliptically polarized and, therefore, is suited to the construction of non-reciprocal ferrite components. The ferric-magnetic column support of the circuit is hereby pre-magnetized by an outer static magnetic field perpendicular to the conductor plane. An additional advantage in using this type of circuit is that components to be subsequently introduced in the circuit, such as diodes and transistors, are used in parallel connection without needing to drill through the usually very hard and brittle substrate. The fields of the slot line can be coupled to those of the microstrip line. Therefore, the slot line is also used in the case of a two-sided utilization of the column support with a microstrip line on one side and, on the other side, with a slot line perpendicular to the direction of the microstrip line at the transfer substation.

In the following section, we will deal with the construction of some components in the stripline technology often used in microstrip design.

## 4.2 STRIPLINE COMPONENTS

In stripline technology, the construction of space and weight saving microwave circuits is possible. These are relatively cheap and reproduce well. In this technology, the circuits are mounted similarly to the well known process of the depressed laminate circuits on a non-conducting carrier plate, the so-called substrate. We usually use the asymmetrical stripline, the so-called microstrip line, because of its simpler structure. With microstrip, the dielectric column support is complete on one side (carrier plate) and with the conductor structures on the other side covered with a well conducting metal. The necessary conductor thickness amounts to at least 3 to 5 times the depth of penetration. Waves propagating in such a circuit (see Ch. 1) are essentially concentrated on the dielectric and can be approximately considered as TEM-waves. A smaller portion of the lines of force runs as stray fields from the edge of the conductor path outward, partially through the bordering medium-air and through the dielectric column support (Fig. 1.26b). This (and the finite conductivity of the conductor material) allows the fields of the stripline to maintain small components in the direction of propagation, and therefore no pure TEM-waves propagate. The length of the stray fields decreases with larger values of the specific inductivities $\epsilon_r$ of the substrate. High $\epsilon_r$ values make a corresponding reduction of the circuit dimensions possible as a result of the decrease in the circuit wavelength ($\lambda_L = \lambda_0/\sqrt{\epsilon_{r_{eff}}}$ . The column support should be low-loss in the interests of a negligible attenuation. In addition to the values of $\epsilon_r$ and the tangent of the loss angle $\delta$, there are additional features in the choice of the substrate for striplines: flat surfaces, mechanical and thermal stability, manageability, and chemical resistance.

Table 4.1 specifies the specific inductivity $\epsilon_r$ for some substrates and the tangent of the loss angle $\delta$ at $f = 10$ GHz and $T = 25°C$. Some 99.5 percent aluminum oxide-ceramic is very often used as a substrate (plate thickness is usually 25 mil = 0.635 mm). Beryllium oxide (which in dust form is poisonous!) can be used if a high heat conductivity is necessary to eliminate the heat due to

**Table 4.1 Column Supports (Substrates)**

| Material | $\varepsilon_r$ | $\tan\delta \cdot 10^4$ at 10 GHz and 25°C |
|---|---|---|
| *a) amorphous* | | |
| polystyrene | 2,54 | 5 |
| teflon | 2,1 | 4 |
| teflon-ceramic powder | 10 ··· 15 | 10 |
| *b) polycrystalline* | | |
| $Al_2O_3$ ceramic, 99.5 percent | 9 ··· 10 | 3 |
| BeO-ceramic, 99.5 percent | 6 ··· 6,6 | 3 |
| quartz | 3,75 ··· 4 | 1 |
| ferrite | 9 ··· 16 | 2 ··· 10 |
| *c) monocrystalline* | | |
| sapphire (monocrystalline $Al_2O_3$) | 9,3 ··· 11,7 | 1 |
| Si (highly resistive) | 11,9 | 4 |
| GaAs (highly resistive) | 12,9 | 4 |

conductor path energy losses. For the construction of gyromagnetic components, we use polycrystalline ferrite materials. Silicon and especially gallium arsenide are used as substrate materials in monolithic circuits.

Stripline circuits are manufactured in thick-film and, more often, in thin-film technology. Striplines are principally used in the frequency range of approximately 3-30 GHz. However, stripline circuits with quartz substrates have already been realized at 140 GHz.

### 4.2.1 Resistors, Capacitors, Inductors

In addition to the use of striplines in microwave circuits as connecting elements, in the realization of reactances in circuit form (filters, matching elements), and as transformation elements and directional couplers, we find integrated resistors, concentrated capacitors and inductors, semiconductors, and ferrite components in such circuits.

In the following section, we will first deal with some *concentrated* passive stripline integrated circuit components which are displayed in Fig. 4.1.

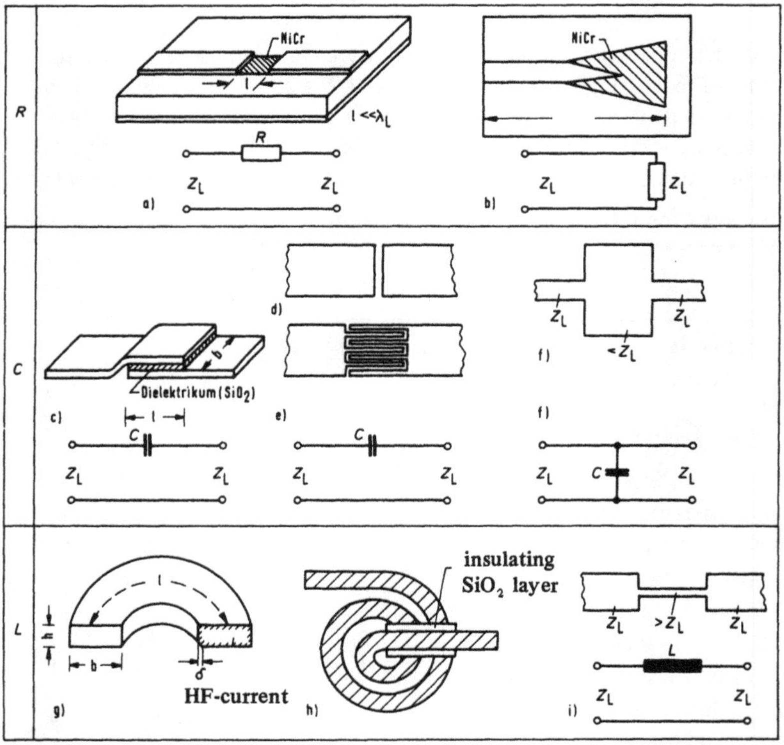


**Fig. 4.1** "Concentrated" strip line component parts

Active concentrated microwave impedances can also often be more simply represented by a short section of a TEM-waveguide. In the longitudinal branch, the equivalent circuit for a section of the differential length d$z$ (Fig. 1.12) contains the resistance $R' \cdot \mathrm{d}z$, the inductivity $L' \cdot \mathrm{d}z$; and, in the cross branch, the shunt conductance $G' \cdot \mathrm{d}z$ and the capacitance $C' \cdot \mathrm{d}z$. The input impedance $\underline{Z}_1$ for such a line of length $l$, which is closed with the impedance $\underline{Z}_2$ amounts to [Eq. (1.111)]

$$\underline{Z}_1 = \underline{Z}_L \frac{\underline{Z}_2 \cdot \cosh \gamma l + \underline{Z}_L \cdot \sinh \gamma l}{\underline{Z}_L \cdot \cosh \gamma l + \underline{Z}_2 \cdot \sinh \gamma l} \tag{4.1}$$

with the line oscillation resistance

$$\underline{Z}_L = \sqrt{\frac{R' + \mathrm{j}\omega L'}{G' + \mathrm{j}\omega C'}} \tag{4.2}$$

and the transmission constants

$$\gamma = \sqrt{(R' + \mathrm{j}\omega L')(G' + \mathrm{j}\omega C')} \,. \tag{4.3}$$

For a very short section of the line ($|\gamma| \cdot l \ll 1$), which is short circuited at the end ($Z_2 = 0$), we obtain with Eqs. (4.1) to (4.3) the approximation

$$\underline{Z}_1 \approx \underline{Z}_L \cdot \tanh \gamma l \approx \underline{Z}_L \cdot \gamma l \approx R' \cdot l + \mathrm{j}\omega L' \cdot l \,. \tag{4.4}$$

we obtain an *inductor* with the $Q$-factor

$$Q = \frac{\omega L'}{R'} \,. \tag{4.5}$$

For a *band-shaped* conductor of width $b$ and height $h$, we obtain for the resistance per unit of length

$$R' = \frac{k \cdot R_\delta}{U} \,. \tag{4.6}$$

Here $U$ is the circumference of the conductor ($\approx 2b$ for $b \gg h$) and $k$ is the corrective factor dependent on the relationship $b/h$, which takes into account the current suppression on the outer region of the conductor (usually $k = 1.3$–$2.0$); $R_\delta$ is the surface resistance, which for copper is $2.61 \cdot 10^{-7} \sqrt{f}$. For the inductivity per unit of length $l$ of level plane conductor strips with $b \gg h$, the following is valid:

$$\frac{L}{\mathrm{nH/cm}} = 2 \cdot \left( \ln \frac{l}{b} + 0{,}2235 \frac{b}{l} + 1{,}193 \right) . \tag{4.7}$$

They are used in the realization of inductivity values of approximately 0.5–4 nH.

Higher inductivity values can be attained by a helical conductor; for a flat spiral, the following is valid:

$$\frac{L}{\text{nH/cm}} = 393\ \text{n}^2 \cdot \frac{\left(\frac{a}{\text{cm}}\right)^2}{8\frac{a}{\text{cm}} + 11\frac{c}{\text{cm}}} \tag{4.8}$$

with

$$a = \frac{d_a + d_i}{4} \quad \text{and} \quad c = \frac{d_a - d_i}{2},$$

$n$ = number of turns, $d_a$ and $d_i$ outer and inner diameters ($d_a/d_i = 0.2$ for maximum efficiency).

Formulas (4.7) and (4.8) are valid for inductivities without screening (distance of the carrier plate from the substrate $> 20b$ and distance from the conductors in the same plane $> 5b$). Under the influence of the carrier plate the inductivity becomes smaller; then, the following is approximately valid:

$$\frac{L_{eff}}{\text{nH}} \approx \frac{\frac{Z_0}{\Omega} \cdot \sqrt{\varepsilon_{r_{eff}}} \cdot \frac{l}{\text{cm}}}{30}, \quad \text{if} \quad Z_0 < \frac{300\ \Omega}{\sqrt{\varepsilon_{r_{eff}}}} \tag{4.9}$$

$Z_L$ = oscillation resistance of this line.

An equivalent circuit of a microwave *resistor* is obtained if we also consider a very short-circuit component and hence ignore $L'$ and $G'$. We then obtain

$$\underline{Y}_1 = \frac{1}{\underline{Z}_1} = \frac{1}{R' \cdot l} + \text{j}\frac{\omega C' l}{3} \tag{4.10}$$

therefore, the parallel connection of a resistance $R' \cdot l$ and a capacitance $C' \cdot l/3$. For flat resistive strips, we therefore, obtain a resistance of

$$R' \cdot l = \frac{R_\delta \cdot l}{b}. \tag{4.11}$$

Examination of a very short circuit component ($|\gamma| \cdot l \ll 1$), which idles at the end ($Z_2 \to \infty$), yields for its input impedance

$$\underline{Z}_1 = \underline{Z}_L \frac{1}{\tanh \gamma l} \approx \underline{Z}_L \frac{1}{\gamma l - \frac{1}{3}(\gamma l)^3} \approx \frac{1}{\gamma l}\left[1 + \frac{1}{3}(\gamma l)^3\right] = \tag{4.12}$$

$$= \frac{1}{(G' + \text{j}\omega C')l} + \frac{1}{3}(R' + \text{j}\omega L') \cdot l \approx \frac{1}{3}R' l + \frac{G'}{\omega^2 C'^2 l} + \text{j}\frac{1}{\omega C' l}.$$

The equivalent circuit therefore, comprises a series circuit composed of a condenser and two resistors. The resistance $R'l/3 = 2R_\delta l/3b$ (designations corresponding to Fig. 4.1c) considers the conductor losses in both condenser plates, and the resistance $G'/\omega^2 C'^2 l$ represents the dielectric losses, which are usually specified by the loss angle $\delta$ with

$$\tan\delta = \frac{1}{Q_d} = \frac{G'}{\omega' C'} \tag{4.13}$$

The reciprocal value of the resulting condenser $Q$-factor is

$$\frac{1}{Q} = \frac{\dfrac{R'l}{3} + \dfrac{G'}{\omega^2 C'^2 l}}{1/\omega C' l} = \frac{R'l + \omega C' l}{3} + \frac{G'}{\omega C'} = \frac{1}{Q_l} + \frac{1}{Q_d}\,. \tag{4.14}$$

In the relations examined above for the ohmic, inductive, and capacitative circuit components, the influence of the surrounding parasitic influences occurring in practice is not taken into consideration.

In Fig. 4.1, some examples of passive concentrated stripline components are assembled. Figure 4.1a shows an integrated series resistance, for which the NiCr adhesive layer is used as the resistive material. The resistance layer length $l$ must be small in comparison to the line wavelength. Resistances up to approximately 200Ω can be easily realized. In the same manner, the circuit termination is produced according to Fig. 4.1b; a low-reflection, wideband circuit termination is attained by a sharp protruding conductor end and a gradually expanding resistive layer (wave basin).

In Figs. 4.1c, d, e, three type models of integrated series condensers (coupling condensers) are outlined. The condenser represented in Fig. 4.1c is formed by the superposition of two conductor path ends separated by a very thin $SiO_2$ layer (thickness approximately 0.5–1μm). The condenser is also designated as an MIM-condenser (metal-insulator-metal); with it we obtain relatively high capacity values of approximately 1–50 pF (100 pF with barium titanite as dielectric). The capacitance according to Fig. 4.1d is formed by the narrow interruption of a conductor path; higher capacitances of approximately 0.1–5 pF can be attained with the so-called interdigital condenser, in which the conductor path ends are formed by narrow fingers gripped into each other.

Reactances can also be realized with an alteration of the oscillation resistance in the stripline by a sudden enlargement or narrowing of the stripline width. This can be easily understood with the help of the relation $Z_L = \sqrt{(L'/C')}$ for the oscillation resistance of a TEM-circuit; a decrease in the oscillation resistance (through an increase in the conductor path width) causes a preponderance of the capacitative effect; whereas an increase in the oscillation resistance increases the inductance influence. In Fig. 4.1f, i corresponding designs of conductor paths are outlined (shunt capacity and series inductivity). Approximation formulas to calculate reactances can be found in sec. 4.2.3 on stripline filters [Eqs. (4.17), (4.18), and (4.19)] and in Fig. 4.4.

Concentrated integrated inductivities can be designed as straight conductor strips; or, in order to obtain higher inductivity values (up to approximately 10 nH) with larger lengths, as flat tuning coils, rectangular coils, or spiral coils (Fig. 4.1g, h). Thin wires (e.g., chip bonding wires) or the vane-shaped beam-lead diode terminals are also illustrated as inductive elements.

With reactances as described above, we obtain $Q$-factors up to approximately 200.

Dummy elements can also be realized as *distributed (line-) elements.* As a susceptance operating transverse to the line, series or parallel resonant circuits can be used at the end of idling or short-circuited tie lines of corresponding length $l$, as Fig. 4.2 shows. At a series resonance ($l = \lambda/4$ with idling tie line, $l = \lambda/2$ with short-circuited tie line) a quasi-short circuit is transformed in the main line as a shunt load for the high-frequency signal, and at a parallel resonance ($l = \lambda/2$ with idling and $l = \lambda/4$ with short circuit) an idling power. The idling power transformed in the short-circuited tie line can be drawn up, for example, for the manufacture of a direct-current return path (HF-throttle).

For the calculation, Eqs. (1.32) and (1.33) are (approximately) valid. We must keep in mind, however, that considering a length measured up to the center line of the main line, small adjustments are necessary. Also, an open-ended microstrip dummy line possesses a complex load at the open end: a real part $G$ as a result of the radiation in the primary mode and an imaginary component $B$ as a result of the energy received in higher power modes. Therefore, and also as a result of the distributed losses in the dummy line, its input conductance is also complex.

Reactances serve, among other things, in the construction of resonators, filters, and matching networks.

We also often use $\lambda/4$ transformation lines as matching elements.

### 4.2.2 Stripline Resonators

Simple stripline resonators can be constructed as circuit components of length $l \approx n\lambda_L/2$, $n = 1, 2, \ldots$, with an idling or short circuit (or reactance) at the circuit's termination. The arrangement of the narrow rectangular strip represents such a *circuit resonator* (fields constant in the $x$-direction) with a line resonance of the $E_{001}$-mode, which means with a local semicycle of the

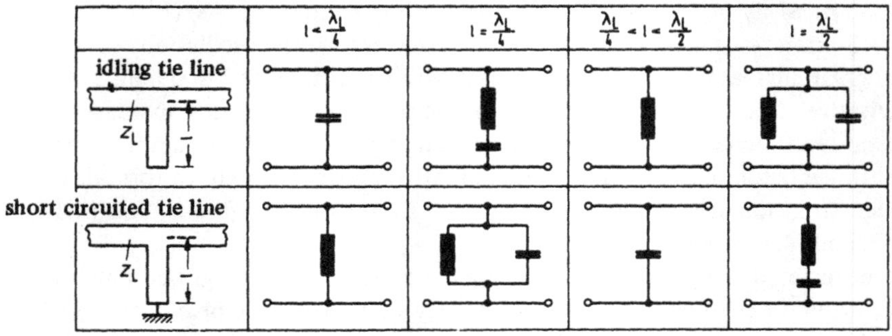


**Fig. 4.2** Tie lines ("distributed" elements)

field distribution in the $z$-direction. In the case of a larger width $b$ of the strip, standing waves can also develop in this direction; then $b \approx m \cdot \lambda_L/2$, $m = 1, 2, 3, \ldots$. In Fig. 4.3, such a *rectangular sheet-resonator* is represented with an $E_{101}$-sheet resonance; therefore, with each there is a local semicycle of the field distribution in the $x$- and $z$-directions.

In Figs. 4.3c, d, e various sheet resonances are represented with their field patterns for a *circular sheet-resonator:* in Fig. 4.3c the fundamental oscillation mode and in Figs. 4.3d, e with their following resonant frequencies the $E_{210}$- and $E_{010}$-modes. The electrical and magnetic fields drawn in Fig. 4.3 are respectively shifted in phase by 90° with respect to each other.

Resonator coupling can take place over the striplines which form a narrow coupling slot with the resonators (electrical coupling). In Fig. 4.4f, a *circular-ring resonator* is outlined with coupling-in and coupling-out lines. Resonance occurs at

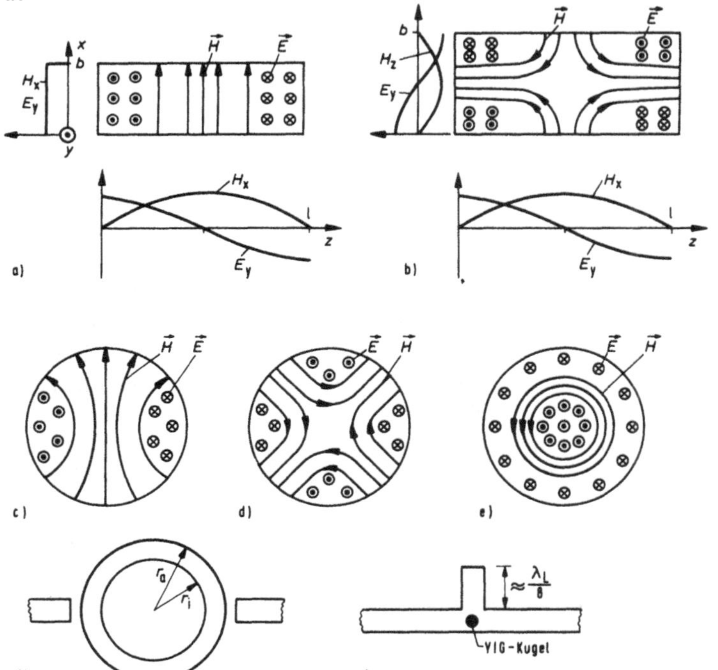


**Fig. 4.3** Resonators: (a) circuit resonator with $E_{001}$-resonance, (b) rectangular sheet-resonator with $E_{101}$-resonance, (c) circular sheet-resonator with $E_{110}$-fundamental oscillation mode, (d) with $E_{210}$-resonance, (e) with $E_{010}$-resonance and (g) ferrite resonator

$$\frac{n \cdot c}{f_n \cdot \sqrt{\varepsilon_{r_{\text{eff}}}}} = 2\pi r_m \tag{4.15}$$

with $r_m$ = mean circle radius.

The $Q$-factor of microstrip resonators is relatively small; the unloaded $Q$ of unloaded resonators is approximately 100 to 1000. Higher resonant $Q$-factors can be attained with triplate or High-Q triplate striplines (Fig. 1.26c, d) than with microstrip lines.

In addition to stripline resonators, *dielectric resonators* are also suitable for integrated circuits, with which we can obtain high $Q$-factors. We use dielectric material, for example, in the form of a rectangular strip or a circular disk. At high $\epsilon_r$ values and low dielectric losses, we obtain unloaded $Q$-factors up to approximately $Q_0 = 50000$. We use this, for example, in oscillation circuits (usually with a FET) in order to improve frequency stability and to reduce FM noise (see also *Mikrowellentechnik,* vol. 2, sec. 3.84).

Very high $Q$-factors up to approximately $Q_0 = 10000$ can also be obtained with circular premagnetized ferrite resonators, the YIG-bullets (ferrite monocrystal, Ch. 5). Figure 4.4 shows an example of such a resonator circuit.

### 4.2.3 Stripline Filters

One field of application for the reactances and resonators is in the construction of microwave filters. Their design can take place according to the familiar process for low-frequency filters, whereby the necessary capacitance and inductance values are realized with the corresponding microwave circuits.

As a simple *low-pass filter,* the arrangements in Fig. 4.4a, b can often be used with realization of the dummy elements through sudden changes in the oscillation resistance of the stripline up to higher values by a line path change and lower values by a widening of the strip conductor. Such low-pass arrangements are used, for example, in matching circuits for semiconductor components. Approximately valid equivalent circuits and equations are specified in Fig. 4.4. The simple approximation formulas can be obtained with the assumption of TEM-wave propagation and with the short line length in comparison to wavelength.

In order to perform the calculations, we examine the input impedance $\underline{Z}_1$ of a non-dissipative circuit with a short length $l \ll \lambda$ compared to wavelength, which is loaded at the end with the impedance $\underline{Z}_2$. The following is valid:

$$\underline{Z}_1 = \frac{\underline{Z}_2 \cdot \cos\beta l + jZ_L \cdot \sin\beta l}{\cos\beta l + j\dfrac{\underline{Z}_2}{Z_L} \cdot \sin\beta l} \approx \frac{\underline{Z}_2 + jZ_L \cdot \dfrac{2\pi l}{\lambda}}{1 + j\dfrac{\underline{Z}_2}{Z_L} \cdot \dfrac{2\pi l}{\lambda}} \approx \tag{4.16}$$

$$\approx \left(\underline{Z}_2 + jZ_L \frac{2\pi l}{\lambda}\right) \cdot \left(1 - j\frac{\underline{Z}_2}{Z_L} \cdot \frac{2\pi l}{\lambda}\right) =$$

$$= \underline{Z}_2 + jZ_L \cdot \frac{2\pi l}{\lambda} - j\frac{Z_2^2}{Z_L} \cdot \frac{2\pi l}{\lambda} + \underline{Z}_2 \cdot \left(\frac{2\pi l}{\lambda}\right)^2 = R + jX$$

and, therefore,

$$X = Z_L \cdot \frac{2\pi l}{\lambda} \cdot \left(1 - \frac{Z_2^2}{Z_L^2}\right). \tag{4.17}$$

The specified mathematical description for an oscillation resistance interruption caused by an abrupt change of the strip width $b$ can be improved by additional consideration of the capacitance $C$ contingent upon the stray electrical field, and of the inductivity $L$ as a result of the concentration of the current density distribution at the contact point (Fig. 4.4c). According to [7] the following are valid:

$$C = \frac{1}{c \cdot Z_{L_1}} \cdot \sqrt{\varepsilon_{r_{eff\,1}}} \cdot \left(1 - \frac{b_2}{b_1}\right) \cdot 0{,}412\,h \cdot \frac{\varepsilon_{r_{eff\,1}} + 0{,}30}{\varepsilon_{r_{eff\,1}} - 0{,}258} \cdot \frac{b_1 + 0{,}262\,h}{b_1 + 0{,}813\,h} \tag{4.18}$$

$$L = \frac{\pi}{4}\,\mu_0 h \cdot \left(1 - \frac{Z_{L_1} \cdot \sqrt{\varepsilon_{r_{eff\,1}}}}{Z_{L_2} \cdot \sqrt{\varepsilon_{r_{eff\,2}}}}\right)^2. \tag{4.19}$$

Figure 4.5a shows a low-pass filter with many such basic parts. In Fig. 4.5b, a low-pass (Cauer low-pass) filter steeped through poles is represented, which is produced in the cross branches with series-resonant circuits. At frequencies above the band-pass width of the filter, the power directed at the filter input is reflected.

Figure 4.5c shows how high-pass behavior can be attained through the use of a circulator (Ch. 5) and a low-pass filter. The charge directed toward the circulator at gate 1 is transferred to gate 2. This charge is reflected at the filter in-

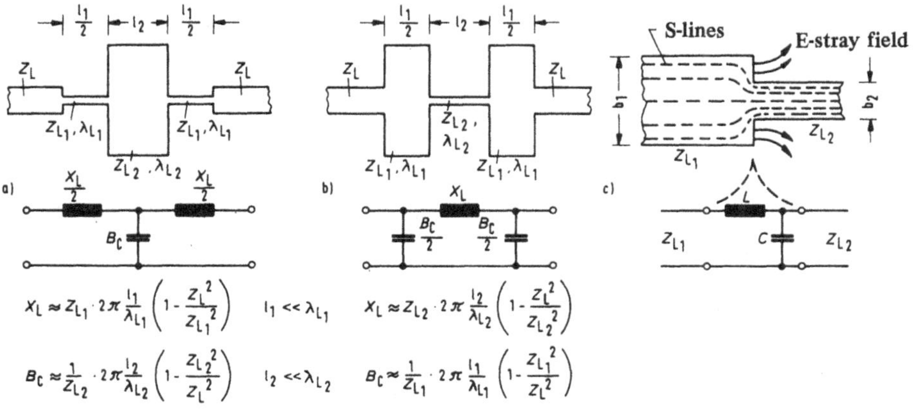


**Fig. 4.4** Reactances realized through oscillatory resistance breaks with approximation formulas

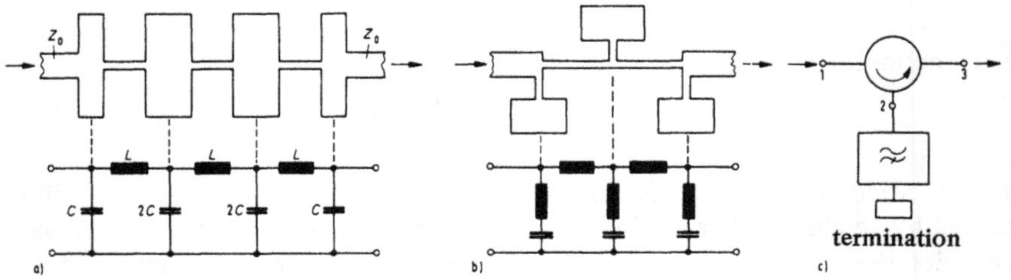


**Fig. 4.5** (a), (b) low-pass filters, (c) realization of a high pass through the use of a low pass and a circulator

put within its suppression range and directed to exit 3 of the circulator. Therefore, the arrangement functions as a high-pass filter.

Examples of *band-pass filters* in stripline technology are outlined in Fig. 4.6. In the simple circuit according to Fig. 4.6a, a line resonator of length $l \approx n \cdot \lambda_L/2$ is arranged between the lines to be connected through at resonance. The center frequencies are approximately

$$f \approx \frac{n \cdot c}{2 \cdot l \cdot \sqrt{\varepsilon_{r_{eff}}}}, \quad n = 1, 2, 3 \ldots \tag{4.20}$$

The filter characteristics can be improved by the use of many coupled resonators. Figure 4.6b shows an arrangement with two stripline resonators coupled at the end. The lateral conductor strips can also be connected directly to the main line as idling tie lines of length $l \approx \lambda_L/2$. With the parallel coupled arrangement of the resonant strips according to Fig. 4.6c, the length of the filter can be reduced and wider cracks between the resonators are permissible. At resonance, a connecting-through extension takes place through the fields excited in the resonators. The line bandpass according to Fig. 4.6d employs capacitative shunt loadings at distances of approximately $\lambda_L/4$, whose effects cancel out at resonance.

With the versions of *band-elimination filters* in stripline design, according to Figs. 4.7a, b, c, $\lambda/2$ resonant circuits are capacitatively coupled to the main line. At resonance, stronger fields are excited in these resonators; in the main line low-impedance resistances are thereby transformed, and the wave fed into the main line is reflected due to the resulting mismatch. Therefore, such filters are also called rejection filters. The bandwidth of the filter can be increased by a gradation of the circuit resonant frequencies. Figure 4.7 shows how a band-elimination behavior can be produced with the help of a bandpass filter and a circulator.

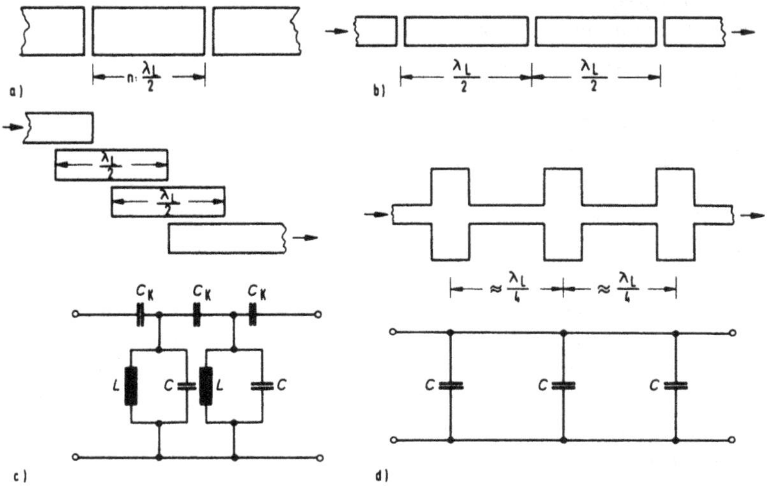


**Fig. 4.6** Band-pass filters: (a), (b), (c) resonance filter coupled to a line, (d) with shunt capacities

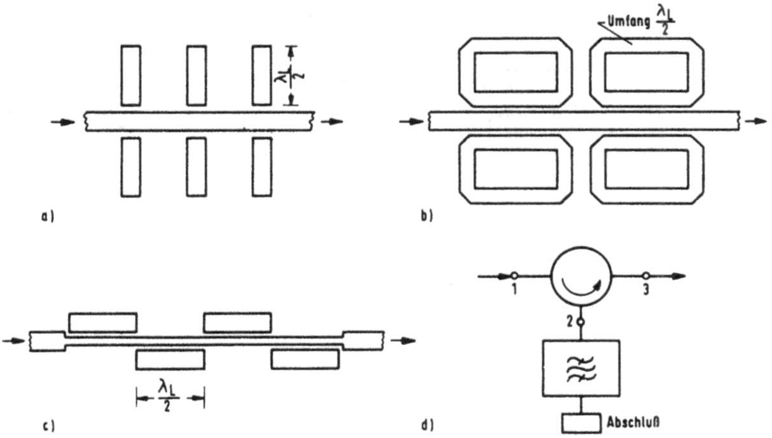


**Fig. 4.7** Band-elimination filters: (a), (b), (c) with coupled λ/2-resonance circuits; (d) with band-pass and circulator filter

With the simple arrangement of coupled striplines, according to Fig. 4.8a, low-pass, band-pass, all-pass, or all-step filters can be realized, depending on which gate is used as an input or output, and whether the other gates are operated open, short-circuited, or connected [30]. Figure 4.8b, c shows two band-pass/band-elimination filters according to the principle of directional couplers. In the case of the arrangement of Fig. 4.8a, lines 1-2 and 3-4 are coupled over a ring with a $\lambda_L$ circumference. Gate 3 is adapted closed. At resonance, a wave fed in arm 1 excites a wave propagating in the clockwise direction within the loop; the second directional coupler couples the wave to arm 4. Therefore, at resonance, transmission takes place toward gate 4 (band-pass filter) and outside of the resonant range toward gate 2 (band-stop filter). In the filter of Fig. 4.8c, the coupling of both lines takes place over two resonant strips, whereby the distance between coupling points on line 3-4 is larger than on line 1-2 by $\lambda_L/2$, so that the couplings occur over the two strips with reversed signs. We obtain a higher selectivity for these filters if we arrange several loops or resonant strips behind each other between the lines.

In filter construction if we use the low-loss triplate stripline (Fig. 1.26c) or the High-Q triplate stripline (Fig. 1.26d), instead of the microstrip line, then steep filter flanks and low transmission losses can be obtained.

In the design of microwave filters, the familiar method for low-frequency filter synthesis (image-parameter method, operating-parameter method, derivation of high- and band-pass filters from low pass filters with the help of the frequency transformation) is used.

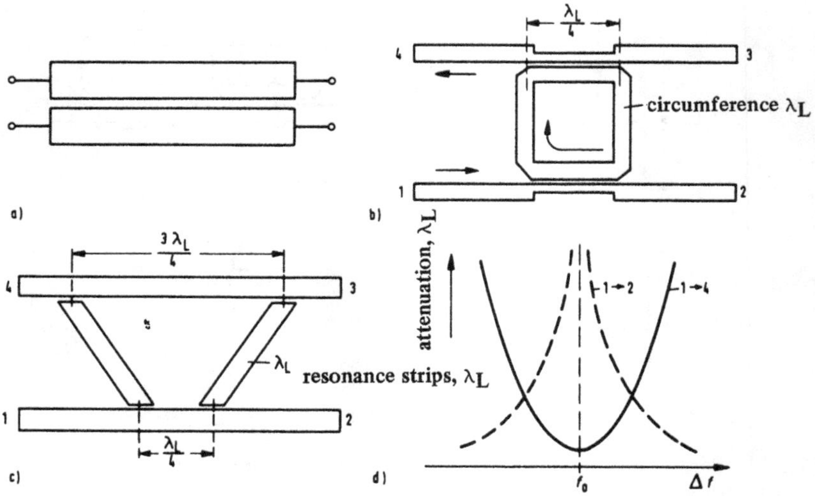


**Fig. 4.8** (a) coupled strip lines as filters, (b), (c) direction filters, (d) attenuation curve to (b), (c)

### 4.2.4 Directional Couplers

An often-used component in stripline technology is the directional coupler (see Ch. 3). Figure 4.9a shows a type model of such a line coupler. Part of the power fed into the main line 1-2 is transferred to the secondary line 3-4 which is directed along a length $l$ parallel to it. Both lines are electromagnetically coupled over the stray fields. Since these lines should only be coupled along the length $l$, they are provided with low-reflection kinks on both sides. In the case of a coupling length

$$l = (2n + 1) \cdot \lambda/4, \qquad n = 0, 1, 2, \ldots \tag{4.21}$$

The part that is proportional to the primary wave in line 1-2 is transmitted to gate 4 and the part proportional to the reflected wave to gate 3 (backward coupler). With such couplers we obtain directional attenuations up to approximately 20 dB.

With this arrangement of strappings lying next to each other (narrow-side couplings), the coupling attenuations are approximately $a_k \geqslant 6$ dB, because otherwise the distance to the strip conductor becomes too small. For more narrow couplings, the conductor paths can be arranged on top of each other (broad-side coupling) with a thin $SiO_2$-intermediate separating layer. Larger bandwidths can be obtained with the arrangement of several coupling sections or an unevenly distributed coupling over a given length (e.g., exponential increase in the conductor distance and breadth).

For calculation of the line coupler, we do not consider the two lines as separated from each other in the coupling range of the waves, but rather we define a new and common waveguide which conducts the superposition of a *symmetrical wave* (even mode) and a *push-pull wave* (odd mode) (Fig. 4.9b). In the case of the even mode, both conductors have the same potential with respect to the conducting plane, and with the odd mode equal potential in the opposite sense.

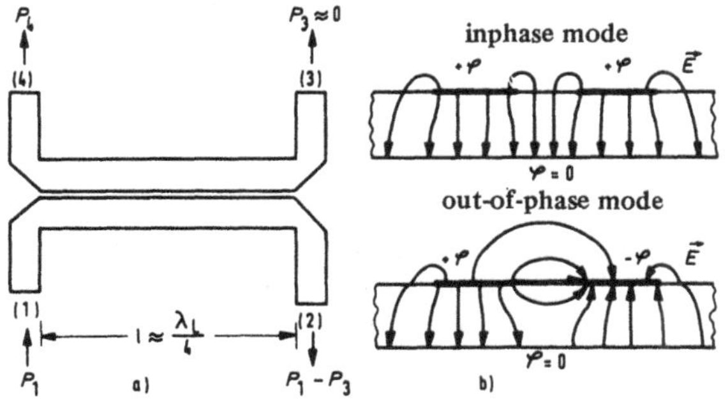


**Fig. 4.9** (a) directional coupler, (b) analysis in inphase and out-of-phase modes

For the phase velocities of the symmetrical wave $v_{p_e}$ and the push-pull wave $v_{p_o}$, the following is valid in comparison with the phase velocity $v_p$ on a simple microstrip line of equal breadth, $v_{p_e} < v_p < v_{p_o}$, since with the push-pull wave more field energy is lost in the air section than with the symmetrical wave. For the oscillatory resistances, $Z_{L_e} > Z_L > Z_{L_o}$ is valid.

### 4.2.5 Hybrid Couplers (Hybrid Bypasses)

Type models of directly coupled (galvanized coupling) directional couplers, so-called hybrid couplers (or hybrid bypasses), in stripline technology are represented in Fig. 4.10. In Fig. 4.10a, a *square-hybrid* (square-fork, two-branch line coupler) is represented (this arrangement is also manufactured in a round form); Fig. 4.10b shows an *H-line coupler* (three-branch line coupler); Fig. 4.10c, a *ring-line coupler* (ring-coupler, ring fork, rat-race coupler); and Fig. 4.10d, a 3-dB ring hybrid (hybrid-ring coupler).

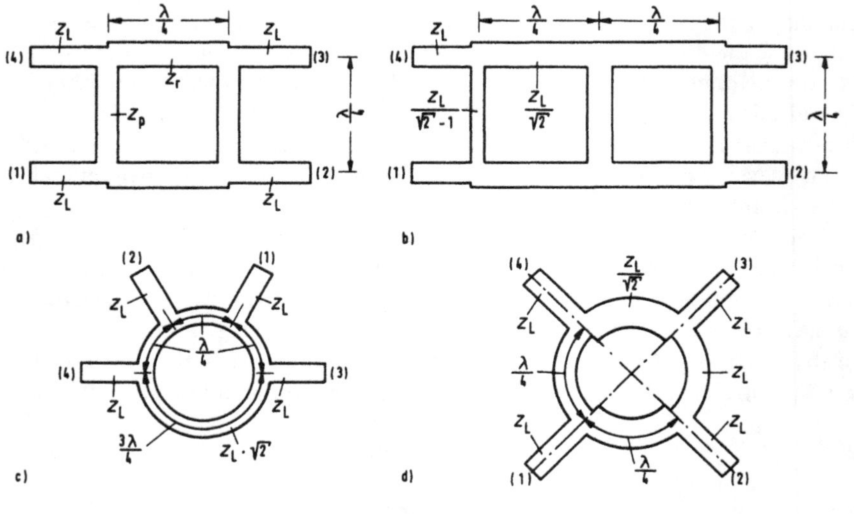


**Fig. 4.10** Hybrid couplers: (a) square hybrid, (b) H-line coupler, (c) ring-line coupler, (d) ring coupler

For the circuits shown, the following is valid: by feeding in gate 1 and a reflection-free termination of the other arms, we obtain a signal distribution to gates 2 and 3, since in each case the partial waves into which the wave fed is divided are in phase. The partial waves in gate 4 cancel out because of the path difference of $\lambda_L/2$, so that gates 1 and 4 are decoupled from each other. Equally, no transmission is possible between gates 2 and 3.

The square-hybrid (Fig. 4.10a) is suitable for couplings of approximately 3–9 dB. For the square-hybrid, the following hold true:

$$\frac{P_2}{P_3} = \left(\frac{Z_L}{Z_{L_p}}\right)^2 , \tag{4.22}$$

$$\left(\frac{Z_L}{Z_{L_r}}\right)^2 = 1 + \left(\frac{Z_L}{Z_{L_p}}\right)^2 . \tag{4.23}$$

For a power distribution in equal parts with the 3-dB coupler, we must choose with $P_2 = P_3 = P_1 \sqrt{2}$

$$Z_{L_p} = Z_L \tag{4.24}$$

and

$$Z_{L_r} = \frac{Z_L}{\sqrt{2}} \quad . \tag{4.25}$$

The signals in gates 2 and 3 possess, as well as in the case of the circuits of corresponding size in Fig. 4.10b, d, a mutual phase difference of 90° (3-dB 90° hybrid). The relative bandwidth of the square-fork is approximately 10–20 percent. An increase in the bandwidth is possible through the introduction of additional cross branches (circuit given in Fig. 4.10b).

The idealized S-matrix of a 90° hybrid coupler, for which all gates are adapted, and gates 1 and 2 as well as gates 2 and 3 are decoupled from each other, reads:

$$\underline{S} = \frac{1}{2} \sqrt{2} \cdot \begin{pmatrix} 0 & \pm j & 1 & 0 \\ \pm j & 0 & 0 & 1 \\ 1 & 0 & 0 & \pm j \\ 0 & 1 & \pm j & 0 \end{pmatrix} . \tag{4.26}$$

For the ring-line fork (Fig. 4.10c) the following is valid: a wave fed in gate 1 divides and moves in the two opposite directions through the ring. Both partial waves in each case are in phase at gates 2 and 3, and out of phase at gate 4. Gates 1 and 4, and gates 2 and 3 are decoupled from each other because of the $\lambda_L/2$ distance. In the design of stripline technology, the arms represent a parallel load for the ring; for the necessary oscillation resistances of the ring and arms, therefore, the following is valid:

$$Z_R = \sqrt{2} \cdot Z_A \, . \tag{4.27}$$

By feeding in gate 1 (*sum* input) and a termination of the other gates, we obtain in-phase output signals at gates 2 and 3, and by feeding in gate 4 (*difference* input) we have out-of-phase signals (0°-180° hybrids) at gates 2 and 3. In the case of a corresponding distance between the gates and the ring, we obtain for the idealized S-matrix [as with the hybrid-T bypass, Eq. (3.26)]

$$S = \frac{1}{2}\sqrt{2}\cdot\begin{pmatrix} 0 & 1 & 1 & 0 \\ 1 & 0 & 0 & 1 \\ 1 & 0 & 0 & -1 \\ 0 & 1 & -1 & 0 \end{pmatrix} \tag{4.28}$$

The ring line bandwidth amounts to approximately 20 percent.

In the waveguide execution of ring-hybrids, according to Fig. 4.10c, the lateral arms represent the series loads for the ring line; now $Z_A = \sqrt{2}\cdot Z_R$ is valid. Sum-and-difference inputs exchange roles.

Applications of hybrid couplers can best be obtained with the connection of two sources to a load, whereby the generators should be decoupled from each other. The decoupling, therefore, takes place in that the signals (fed in the decoupled gates 1 and 4) of both sources in each case pass through a path which differentiates them in length by $\lambda_L/2$. An example of application is the push-pull mixer according to Fig. 3.34, with an input circuit and oscillator circuit decoupled from each other.

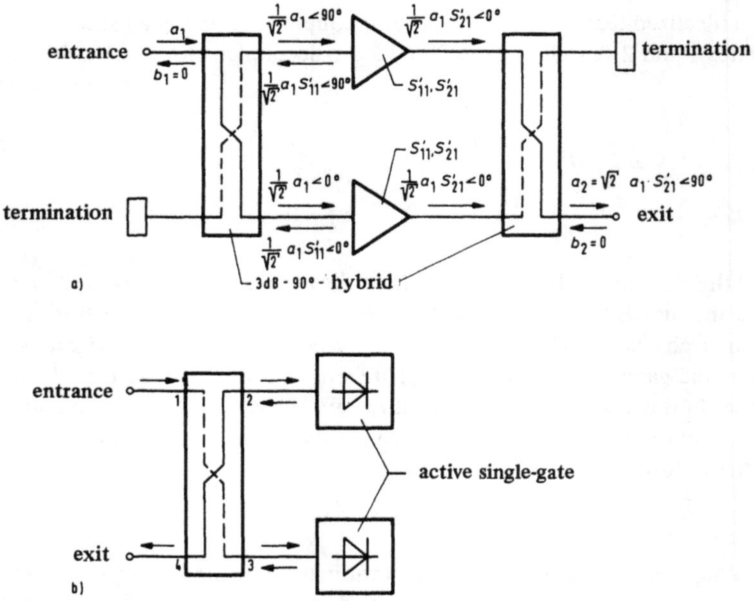


**Fig. 4.11** (a) balance intensifier, (b) with reflection amplifiers

Hybrids can also be used in the improvement of the power standing-wave ratio, if reflecting components must be used in a circuit, Fig. 4.11a (specified relationships with ideal hybrids). The reflected charges are absorbed by the terminal resistances. This circuit arrangement is often used to bring together the charges of two intensifier stages (balance amplifier, introduced by Engel-Brecht, Bell Labs, 1964). In Fig. 4.11b, the circuit principle of a hybrid coupled intensifier with single-gate intensifiers (reflection amplifiers) is represented. The input signal directed to gate 1 of the hybrid is distributed in equal parts to the equally terminated gates 2 and 3; gate 4 is decoupled from gate 1. The signals are intensified in both reflection amplifiers connected to gates 2 and 3, and then reflected back into gates 2 and 3. These reflected waves add in phase at gate 4 and cancel out at gate 1, because of their phase opposition. If both single-gate intensifiers do not agree in the amount and phase of their input impedances, then part of the intensified signal is conducted back to input gate 1. Through the use of additional hybrids as power dividers or combiners, the circuit according to Fig. 4.11a can be enlarged with additional intensifiers. The use of hybrids in symmetrical mixer circuits is shown by Fig. 3.34.

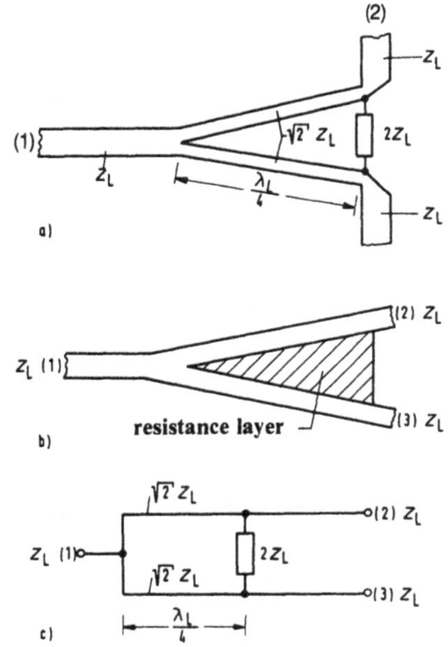


**Fig. 4.12** Hybrid power dividers, (a) with concentrated resistance $2Z_L$, (b) with resistance layer, (c) equivalent circuit of (a)

Hybrids are often used as *power dividers.* Power dividers (parallel bypasses) are most often constructed as resistance-, reactance-, or hybrid-power dividers.

Figure 4.12 shows two triple-gate hybrid power dividers in stripline technology with a power distribution in equal parts (3-dB divider) and in-phase outputs, in Fig. 4.12a with a concentrated resistance $2 \cdot Z_L$, in Fig. 4.12b with resistance strips, and in Fig. 4.12c the equivalent circuit of *b*. With such power dividers we attain bandwidths up to one octave.

An additional important stripline component, the Y-circulator, will be dealt with in Chapter 5 which covers ferrite components.

## 4.3 WAVEGUIDES FOR MILLIMETER WAVE MICs

We designate the frequency range from 30 to 300 GHz ($\lambda$ = 1 to 10 mm) as the millimeter wave range. Primarily resulting from the advances made in the development of semiconductor elements for this range, the use of millimeter waves has sharply increased. Fields of application are the communications transmission (point-to-point connections, short distances, high transmission capacities), radar technology* (short-range radar, search and tracking radars for airborne objects, steering radars for aircraft and missiles, detonators, air traffic control radars, and industrial radars), radiometry for the discovery of ground targets because of its temperature emission, radioastronomy, meteorology, plasma investigation, spectrometry, and medicine. The use of millimeter waves brings with it some advantages: higher bandwidths are possible, sharp directivity with relatively small antenna apertures, high radar resolution (e.g., small targets in a dense clutter environment) and precision, larger Doppler shift ($fd = 2\nu_r/\lambda$, $\nu_r$ = radial velocity), the ability to pierce clouds, snow, fog, smoke, and dust, less probability of interference and detection as opposed to optical and infrared sensors, and above all the possibility of a compact construction with light weight. Preferably, we work at frequencies in the "transmission windows" with low atmospheric attenuation at 35, 94, 140, and 230 GHz (occasionally also at the absorption minima, so as not to be discovered or overheard).

There are numerous waveguides suitable for millimeter wave circuits. Coaxial transmission lines are only suitable for frequencies up to approximately 40 GHz, since at higher frequencies their cross-section dimensions become too small and small cross sections are necessary in order to prevent wave excitation in the hollow waveguide. Standard rectangular hollow waveguides are produced for frequencies up to 330 GHz. By increasing the waveguide height ($b > a$), the waveguide attenuation may be reduced. Such oversized rectangular hollow waveguides are also suitable for even higher frequencies; in any case, higher wave modes can be excited.

Since the production of circuits in waveguide technology is costly, we usually use planar or quasi-planar waveguide structures in the production of economical microwave integrated circuits suitable for mass production. Figure 4.13 gives an overview of some customary hollow waveguides suitable for millimeter waves and used in the construction of millimeter wave MICs.

*C. Rint, *op. cit.*, articles on radar technology, vol. 4, pp. 531-722; and vol. 5, pp. 709-766.

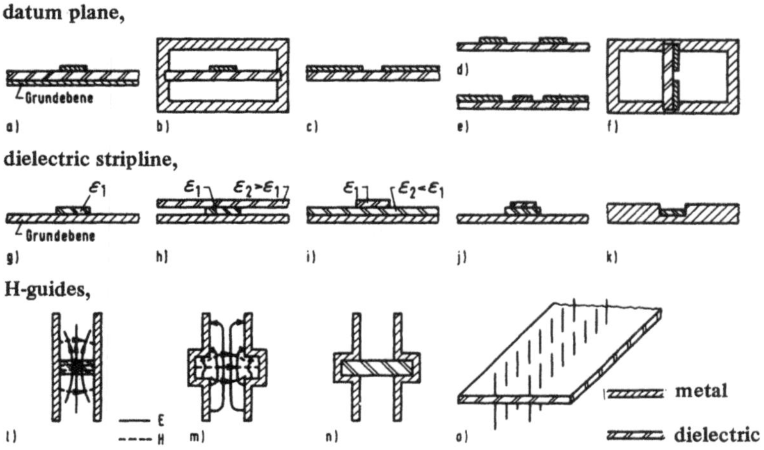


**Fig. 4.13** Hollow waveguides suited to mm-wave MICs (in the cross section): (a) microstrip line, (b) Brenner-line suspended substrate-strip line (c) slot line, (d) double-slot line, (f) tail or fin-line, (g) dielectric image line, (h) inverted strip-, (i) insulated-, (j) modified-, (k) imbedded dielectric image line, (l) H-guide, (m) groove guide, (n) dielectric groove guide, (o) fence line

*Microstrip lines* (Fig. 4.13a) are used for frequencies up to approximately 100 GHz. The limit is set by high losses, the excitation of leaky surface waves propagating from the conductor strips along the substrate, radiation at discontinuities, and tolerance problems due to the manufacturing process. The line losses of the microstrip line are relatively high.

Figure 4.13b shows a shielded stripline, a so-called Brenner-line (after H.E. Brenner), suspended substrate line; it is also used with conductor strips on both sides (bilateral suspended substrate line). The metal casing takes over the task of the metal conducting plane. Since the field is not concentrated on the dielectric column support, as was the case with the microstrip line, the losses are less. The substrate of the microstrip line with the conductor path on the upper side can be mounted on a broad side wall inside of the screening box.

The *slot line* (Fig. 4.13c) and the *coplanar* (or double slot) *line* (Fig. 4.13d) exhibit no (quasi-) TEM-wave propagation (no current paths for longitudinally directed line currents). Since the magnetic field possesses a component in the direction of propagation (Fig. 4.13f), there are areas of elliptical polarization for an incident wave. Therefore, these lines are suitable for the construction of non-reciprocal components using ferrite materials on the substrate side. Another advantage is that with slot and coplanar lines, parallel connecting hybrid components (resistors, condensers, diodes) can be introduced in the circuit without drilling through the usually hard and brittle substrate plate. The field of the slot line can be easily coupled with that of the microstrip line. Thus, in the case of two-sided utilization of the column support, we can simply couple a slot line on one side of the substrate with a microstrip line on the other side at a location of perpendicular wiring. A material with a high $\epsilon_r$ is required for the slot line substrate, so that the fields in the vicinity of the slot are concentrated and the radiation losses remain low.

Shielded slot and coplanar lines, such as that in Fig. 4.13f, are designated as *tail* or *fin-lines.* Fin-lines can also be considered twin-ridge hollow waveguides with a very narrow ridge. Figure 4.13f shows the basic structure of a one-sided, or unilateral, fin-line. In the case of a two-sided, or bilateral, fin-line, such an arrangement is also located on the other substrate side. It shows low losses and more current supply possibilities for semiconductors; the line oscillation resistances are approximately $Z_L \geqslant Z$ 100Ω. The antipodal fin-line possesses only one conductor path on each of the two substrate sides, which border respectively on another broad side. With this fin-line, small oscillation resistances up to approximately 10Ω or less can be realized. Thus, for example, it is suitable as a transition between a rectangular hollow waveguide and a microstrip line. In the case of a coplanar fin-line the conductor arrangement of the coplanar line (double-slot line) is located on the transverse substrate. Fin-lines have (similarly to ridge hollow waveguides) the advantage of a large clear bandwidth (2:1), and therefore allow the manufacture of wideband circuit components. The insertion of semiconductors is easily done and the fin-lines can simply be combined with other line forms.

In the diagrams of Fig. 4.13g-k, several *dielectric striplines* are represented in the cross section. Figure 4.13g shows with the *dielectric image line* (or image waveguide, according to D.D. King, 1955), the basic form of which is composed of a dielectric strip on a metal conducting plane. The designation dielectric image line can be explained by the fact that we obtain through the reflection at the metal plate, a dielectric wave conductor with double the height. A disadvantage with this line is that, in the case of a concentration of the fields on the line (at high $\epsilon_r$), the losses in the dielectric and the conductor plate are high; whereas in the case of low losses the fields extend deeply into the area.

An improvement can be brought about by mounting a dielectric plate on the dielectric image line in the case of the inverted strip dielectric waveguide (Fig. 4.13h), or with the versions of the insulated image line (Fig. 4.13i) and the modified image line (Fig. 4.13j). At discontinuities and contortions of these lines, a strong radiation takes place. The imbedded dielectric line (trough line or trapped image line, Fig. 4.13k) also belongs to the *H*-waveguide type.

The *H-waveguide* is an open waveguide with an "H"-shaped cross section; Fig. 4.13l shows its basic structure. The field distribution is laterally restricted by metal plates, and limited up and down by surface wave propagation with the help of a dielectric cross plate. The energy is directed along the line in the form of surface waves (hybrid modes, all six field components present), which are reflected between the lateral walls, and the energy is concentrated on the dielectric strips and in its immediate vicinity. The energy decreases exponentially upward and downward with the distance from the plate. Similarly, as in the case of the oversized rectangular hollow waveguide, the attenuation is low. The dielectric strip limits, however, the excitation of higher order modes. Just as with the oversized rectangular hollow waveguide, it can also be used at frequencies greater than 400 GHz (also like the groove waveguide, Fig. 4.13m). The *H*-waveguide exists in many variations: thus, several parallel dielectric strips can be arranged between the metal side walls; we can also attain the field limitation toward the open sides with the help of a metal delay-action structure operating as a synthetic dielectric (see also *Mikrowellentechnik,* Vol. 2, Ch. 1 and 2) in the form of metal cross strips arranged in parallel behind each other in the direction of propagation. Another form is the imbedded dielectric line (trough waveguide), according to Fig. 4.13k. The groove guide (Fig. 4.13m, directs transverse-electrical modes, $H_{11}$-mode in Fig. 4.13m) can also be assigned to the *H*-waveguide structures; it possesses similar characteristics (*H*- and groove waveguide, according to F.J. Tischer, 1952). The bulge moving in the longitudinal direction in the side walls also produces the field formation of a surface wave in its immediate vicinity. In this way the excitation of higher order modes is reduced. Through the dissipation of the dielectric cross strips, the attenuation is even less and similar to the oversized rectangular hollow waveguide. Figure 4.13n shows a dielectric groove waveguide. The fence, or comb line, waveguide, according to Fig. 4.13o, is also an *H*-waveguide form. At the lateral metal plate locations, wire grids are used; the field distribution is similar, and also decreases exponentially above and below the line.

The *H*-waveguides, including the groove and fence waveguides, appear suitable for economical mass production of millimeter wave MICs.

# CHAPTER 5

# FERRITE COMPONENTS (GYROMAGNETIC COMPONENTS)

In this chapter we describe microwave components, whose special characteristics are based on the interaction of magnetic materials with electromagnetic waves. So that such interaction takes place, the wave must be able to penetrate the material deeply enough. This is possible with materials with ferrimagnetic characteristics, ferrites.

In contrast to ferromagnetic materials (Fe, Ni, Co), ferrites possess ceramic characteristics. Since they exhibit very low electrical conductance, their eddy current losses remain small even at very high frequencies, so that they can be used in the microwave range. Ferrites are manufactured as sinter ceramics or as monocrystals. We sinter (at 1000–1500°C) a mixture of metallic oxides (-nitrates, -carbonates), whereby the metallic oxide connections are of the form $MeOFe_2O_3$, where Me is a bivalent metal (Mg, Cd, Mn, Fe, Co, Mi, Cu, Zn). Ferrites are also used with other crystal structures, such as, for example, barrium ferrites (for millimeter waves ferrite parts, because a smaller outer magnetic field is required), or yttrium-iron-garnets (YIG); the YIG-material is usually processed as monocrystalline pellets.

In microwave technology, ferrites are mostly used in the construction of asymmetrical transmission (non-reciprocal) components. The ferrite material maintains the necessarily strong magnetic anisotropy with an outer static magnetic field. The permeability of the ferrite is then no longer dependent on the field direction and becomes a skewed-symmetric tensor. By using the direction-controlled characteristics of premagnetized ferrites, numerous microwave components can be manufactured, such as one-way circuits, circulators, attenuator pads, phase shifters, interrupters, and modulators. The mode of operation can be explained by the precession of the electron spin.

## 5.1 PHYSICAL FUNDAMENTALS

According to the classical model of magnetism, we can conceive of the magnetic characteristics of a material as having arisen from the effect of the electrons spinning on their own axis with a specific torque.

An electron spinning on its axis can be assigned a mechanical angular momentum $\vec{D}$ (spin momentum) because of its mass with

$$D = \frac{\hbar}{4\pi} \tag{5.1}$$

$\hbar = 6{,}63 \cdot 10^{-34}\ \mathrm{Ws}^2$ = Planck's radiation law.

The rotation of the (negative) electrical load produces a magnetic dipole momentum $\vec{m}_B$ (Bohr magneton) parallel to the rotation axis, whereby the direction is a function of the rotation direction and the electron behaves as a magnetic dipole. The gyromagnetic relation is valid:

$$\vec{m}_B = -\Gamma \cdot \vec{D}\,. \tag{5.2}$$

Here,

$$\Gamma = \frac{e}{m} \cdot \mu_0 = 2{,}21 \cdot 10^5 \frac{m}{A \cdot s} \tag{5.3}$$

is the gyromagnetic relationship, $e, m$ = charge and mass of the electron.

The rotation axes of the electrons of an atom are indeed aligned, however, the spin momentum occurs in the atoms of a non-magnetic material in antiparallel pairs, so that the atom has no magnetic momentum. In steady-state ferrimagnetic substances, the magnetic spin momentum is aligned in the same orientation as a result of the strong coupling between the atoms; and, therefore, yields maximum magnetic effect. With ferrimagnetic materials (ferrites), the effect of the coupling is that the atoms appear in two groups with antiparallel aligned electron spins; in this case the number of the spin movements in both groups is equal, so that a magnetic field occurs on the outside.

The magnetic field on the inside of the ferrite probe is different from the magnetic field installed from the outside. The magnetic field intensity $\vec{H}_i$ on the inside of the ferrite is formed by the anisotropy fields of the Weissian zones and a magnetic field eventually installed on the outside. Besides this, demagnetizing factors, corresponding to the geometry of the probe and the direction of the outer magnetic field, are to be taken into consideration.

Every magnetically active electron, which possesses a magnetic momentum $\vec{m}_B$ because of its spin, undergoes a mechanical torque $\Theta$ under the influence of the magnetic field intensity $\vec{H}_i$ on the inside of the ferrite, which is directed perpendicular to $\vec{m}_B$ and $\vec{H}_i$ (Fig. 5.1).

$$\vec{\Theta} = \vec{m}_B \times \vec{H}_i\,.$$

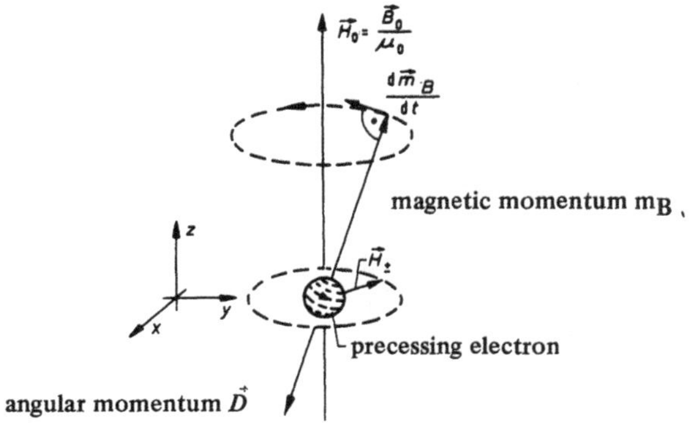


**Fig. 5.1** Electron spin

This momentum and a motion-damping torque $\vec{\Theta}_d$ change the angular momentum $\vec{D}$, for which the following is generally valid:

$$\vec{\Theta} = \frac{d\vec{D}}{dt} \tag{5.5}$$

which corresponds to the differential equation

$$\frac{d\vec{D}}{dt} = \vec{m}_B \times \vec{H}_i + \vec{\Theta}_d \,. \tag{5.6}$$

With Eqs. (5.6) and (5.2), we obtain for the magnetic momentum $\vec{m}_B$, without taking the attenuation into consideration, the dynamic equation

$$\frac{d\vec{m}_B}{dt} = -\Gamma \cdot \vec{m}_B \times \vec{H}_i \tag{5.7}$$

According to this, the vector $\vec{m}_B$ moves in the case of a magnetic constant field $\vec{H}_i = \vec{H}_0$ with the angular velocity

$$\omega = \Gamma \cdot |\vec{H}_0| \tag{5.8}$$

on a cone-shaped shell with an axis parallel to $\vec{H}_0$ (Fig. 5.1). If, therefore, an electron is located in a static magnetic field, then the torque $\Theta$ exerted on the magnetic dipole [Eq. (5.4)] produces a free precession movement of the dipole axis (electron rotation axis) around the direction of the magnetic field. The vector of the magnetic momentum of the electron spins around the direction of $\vec{H}_0$ in the right-hand direction. The precession angular velocity $\omega$, according to Eq. (5.8), is also designated as the L'Armor frequency. In reality, the precession movement described experiences an attenuation (attenuation momentum [$\Theta_d$ in Eq. (5.6)] through the inner frictional losses in the ferrite material.

Without an outer magnetic field the directions of anisotropic fields inside the ferrite probe are randomly distributed and cancel each other out by their sums. If we attach a magnetic constant field $\vec{H}_0$ onto the material, then the magnetic movements are aligned. As a result of the angular momentum of the spinning electrons, the alignment does not follow directly, but rather as an attenuated precession movement, similar to a rotating gyroscope disturbed in the same manner. As a result of the effective attenuation, the aperture angle of the precession pin, which is described by the magnetic momentum, decreases until $\vec{m}_B$ and $\vec{H}_0$ lie in one direction and the torque $\vec{\Theta} = \vec{m}_B \times \vec{H}_i$ becomes zero. In this case the electron rotation axis (direction of the magnetic momentum) moves on a spiral-shaped path in the direction of the installed magnetic field. The closer $H_0$ approaches the saturation field intensity of the probe, the more the direction of the inner field will agree with that of $\vec{H}_0$. If the magnetic field installed is sufficient for the magnetic saturation, then all effective magnetic momentum is adjusted in the direction of the outer field after a certain amount of time.

If, in addition to the static magnetic field, an alternating magnetic field $\vec{H}$ with a suitable frequency and direction (perpendicular to $\vec{H}_0$) bears upon the electrons, then a forced precession movement can be maintained in the vertical direction. If the alternating magnetic field has such an effect that the vector of the total field intensity $\vec{H}_0 + \vec{H}$ is constantly set at a given angle with respect to the $\vec{H}_0$ direction, then the torque can be increased and, thereby, the attenuation influence can be compensated. The losses transformed into heat grow with an increasing aperture angle of the precession pin, which is adjusted such that the alternating field power input becomes the same as the dissipation loss.

Next, we will more closely consider the case in which a homogeneous static magnetic field $\vec{H}_0$ acts on a ferrite probe, and upon which a homogeneous alternating magnetic field $\vec{H}$ is superimposed. Between the inductive $\vec{B}$ and the magnetic field intensity $\vec{H}$ of the premagnetized ferrite, the following relation is valid:

$$\vec{B} = \|\mu\| \cdot \vec{H} \tag{5.9}$$

with the tensor $\|\mu\|$ of the direction-controlled permeability (so-called Polder tensor, named after D. Polder). In the following, $\|\mu\|$ will be determined by the condition that the homogeneous magnetic field consist of an equal part and a relatively smaller alternating part, and the equal part must be large enough in the

$z$-direction that the ferrite is saturated. These conditions are usually met in the case of microwave ferrite applications. The inner magnetic induction $\vec{B}_i$ of the ferrite is composed of the free space induction $\mu_0 \cdot \vec{H}_i$ and the sum of all the magnetic momentum in the volume $V$ under consideration.

$$\vec{B}_i = \mu_0 \vec{H}_i + \sum_n \frac{\vec{m}_{B_n}}{V} = \mu_0 \cdot \vec{H}_i + \vec{M}_i$$

$$\text{with} \quad \vec{M}_i = \sum_n \frac{\vec{m}_{B_n}}{V} = \text{magnetization} \tag{5.10}$$

The magnetic field intensity $\vec{H}_i$ on the inside of the probe can be calculated, in the case of saturation, from its geometry, the outer field intensity, and the saturation magnetization. If we assume the probe to be spherical in the interests of simplification, then no demagnetized field need be considered (demagnetization factors for all three directions in space equal to $4\pi/3$) and we can assume $\vec{H}_i = \vec{H}_0$.

Since the ferrite is premagnetized into saturation, the electron spins of all magnetically active electrons whose number per volume is $n$, are oriented in the same direction and their magnetic movements are added to the magnetization:

$$\vec{M}_i = \sum_n \frac{\vec{m}_{B_n}}{V} = n \cdot \vec{m}_{B_n} . \tag{5.11}$$

Therefore, this is only capable of changing its direction, not its magnitude. In the magnetically saturated ferrite, there is a steady coupling between the individual electron spins, so that the material operates as a single magnetic dipole.

The dynamic equation (5.7) yields for the phase delay of the magnetization $\vec{M}_i = \vec{B}_i - \mu_0 \vec{H}_i$:

$$\frac{d}{dt}(\vec{B}_i - \mu_0 \vec{H}_i) = -\Gamma(\vec{B}_i - \mu_0 \vec{H}_i) \times \vec{H}_i = -\Gamma \cdot \vec{B}_i \times \vec{H}_i . \tag{5.12}$$

For $\vec{H}_i$ and $\vec{B}_i$, formed by an equal part in the $z$-direction and an alternating part, the following statements are valid:

$$\vec{H}_i = \vec{H}_0 + \vec{H} = \vec{e}_x \cdot H_x + \vec{e}_y \cdot H_y + \vec{e}_z (H_z + H_0) \tag{5.13}$$

$$\vec{B}_i = \vec{B}_0 + \vec{B} = \vec{e}_x \cdot B_x + \vec{e}_y \cdot B_y + \vec{e}_z (B_z + B_0) . \tag{5.14}$$

If we substitute the expressions (5.13) and (5.14) into the relation (5.12) between $\vec{B}_i$ and $\vec{H}_i$, then after completing the cross product and ignoring the small products of the alternating quantities at negligible modulation (small alternating field amplitude $\hat{H} \ll H_0$), obtain in complex form ($d/dt = j\omega$)

$$\begin{aligned}
j\omega\,(B_x - \mu_0 \cdot H_x) &+ \Gamma(B_y \cdot H_0 - H_y \cdot B_0) = 0 \\
j\omega\,(B_y - \mu_0 \cdot H_y) &+ \Gamma(H_x \cdot B_0 - B_x \cdot H_0) = 0 \\
j\omega\,(B_z - \mu_0 \cdot H_z) &= 0\,.
\end{aligned} \tag{5.15}$$

Therefore, for the relation $\vec{B} = \|\mu\| \cdot \vec{H}$ between $\vec{B}$ and $\vec{H}$, we obtain

$$\begin{pmatrix} B_x \\ B_y \\ B_z \end{pmatrix} = \begin{pmatrix} \mu_1 & j\mu_2 & 0 \\ -j\mu_2 & \mu_1 & 0 \\ 0 & 0 & \mu_0 \end{pmatrix} \cdot \begin{pmatrix} H_x \\ H_y \\ H_z \end{pmatrix}, \tag{5.16}$$

whereby

$$\mu_1 = \mu_0 \left(1 + \frac{\omega_0 \cdot \omega_m}{\omega_0^2 \omega^2}\right) \tag{5.17}$$

$$\mu_2 = \mu_0 \cdot \frac{\omega \cdot \omega_m}{\omega_0^2 - \omega^2} \tag{5.18}$$

and

$$\omega_0 = \Gamma \cdot H_0 \tag{5.19}$$

$$\omega_m = \frac{\Gamma}{\mu_0}(B_0 - \mu_0 \cdot H_0) = \frac{\Gamma}{\mu_0} \cdot M_s\,. \tag{5.20}$$

$f_0 = \omega_0/2\pi$ is the *gyromagnetic resonant frequency* and $M_s$ the *saturation magnetization.*

The permeability $\|\mu\|$ of the ferrite is obtained, corresponding to Eq. (5.16), as a tensor (Polder tensor), which means it is direction-controlled; only in the special case of circularly polarized fields does the HF-permeability become scalar again [Eq. (5.26)].

If the frequency $f$ of the excited alternating magnetic field is equal to the gyromagnetic resonant frequency $f_0$ (natural precession frequency of the material), then magnetic resonance takes place with strong interaction between the high-frequency field and the electron spin: according to Eqs. (5.17) and (5.18), $\mu_1, \mu_2 \rightarrow \infty$ and, therefore, $B_x, B_y \rightarrow \infty$. In reality these magnitudes only attain finite maximum values since the precession movement is attenuated (due to reflection effects).

If we consider the attenuation by an attenuation constant $\alpha$, whereby we replace $\omega_0$ with $\omega_0 + j\alpha$ in Eqs. (5.17) and (5.18), then we obtain

$$\mu_1 = \mu_0 \left(1 + \frac{(\omega_0 + j\omega\alpha) \cdot \omega_m}{(\omega_0 + j\omega\alpha)^2 - \omega^2}\right)$$

$$\mu_2 = \mu_0 \cdot \frac{\omega \cdot \omega_m}{(\omega_0 + j\omega\alpha)^2 - \omega^2} \cdot \qquad (5.21)$$

We obtain the attenuation constant as

$$\alpha = \frac{1}{\omega\tau}, \qquad (\alpha \ll 1), \qquad (5.22)$$

whereby $\tau$ is the fade-out time of the free (that is at $H = 0$) electron spin precession (usually $\tau \approx 10^{-8}$ s).

The influence of the direction-controlled permeability on wave propagation in a premagnetized ferrite material can be easily understood by the example of a medium with infinite length. From Maxwell's equations (1.7) and (1.8), upon elimination of the electrical field intensity, we obtain the equation:

$$\text{rot rot } \vec{H} - \omega^2 \cdot \|\mu\| \cdot \varepsilon \cdot \vec{H} = 0 \qquad (5.23)$$

with the tensor $\|\mu\|$ of the direction-controlled permeability. If we examine a wave propagating in the $z$-direction, in which the ferrite is premagnetized, then we obtain as a solution in Cartesian coordinates

$$\vec{H}_\pm = (\vec{e}_x \mp j\vec{e}_y) \cdot \hat{H} \cdot e^{j\omega t - \gamma_\pm \cdot z} \qquad (5.24)$$

with

$$\gamma_\pm = \alpha_\pm + j\beta_\pm = j\omega \cdot \sqrt{\mu_\pm \cdot \varepsilon} \qquad (5.25)$$

and

$$\mu_{\pm} = \mu'_{\pm} - j\mu''_{\pm} = \mu_1 \pm \mu_2 = \left(1 + \frac{\omega_m}{\omega_0 \mp \omega + j\omega\alpha}\right) \cdot \mu_0 \,. \qquad (5.26)$$

Therefore, two wave modes with different propagation constants $\gamma_+$ and $\gamma_-$ are capable of propagation. Since the x- and y-components of $\vec{H}_+$ and $\vec{H}_-$ are equal in magnitude according to Eq. (5.24), although deplaced in phase by $\pm\,\pi/2$, $\vec{H}_+$ and $\vec{H}_-$ are circularly polarized parallel to the xy-plane, which means perpendicular to the $H_0$-field showing in the z-direction. The direction of $\vec{H}_+ = (\vec{e}_x \cdot H_+ \cdot e^{j\omega t} + \vec{e}_y \cdot H_+ \cdot e^{j(\omega t - \pi/2)}) \cdot e^{\gamma_+ \cdot z}$ rotates around the $z$-axis (direction of the static premagnetization field) in a mathematically positive direction, which means in the direction of the increasing pole angle values $\varphi$ (examine, for example, $H_{+x} = H_+ \cdot \cos(\omega t - \gamma_+ \cdot z)$ and $H_{+y} = \sin(\omega t - \gamma_+ \cdot z)$ for $t = 0$ and $t = T/4$). We call such a wave, for which the field vectors seen in the direction of propagation ($H_0$ direction) rotate in time in the clockwise direction at a fixed location, as a *clockwise* or *positive circularly polarized wave.* Correspondingly, the direction of $H_-$ rotates in a mathematically negative direction around the $z$-axis; the wave is *counterclockwise* or *negative circularly polarized.* In the complex $(x + jy)$ plane, we obtain the rotational vector $\underline{H}_{\pm} = H \cdot e^{\pm j\omega t}$. Thus, $H_+$ is clockwise as viewed from the $H_0$ direction and $H_-$, counterclockwise.

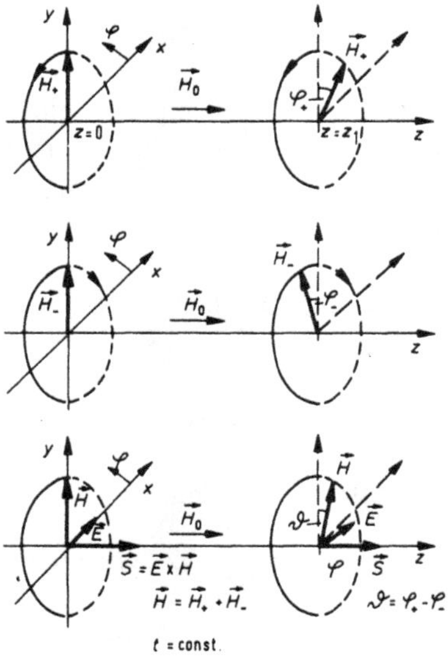


**Fig. 5.2** Rotation of the polarization direction at the angle $\theta$ at the distance $z$, according to the Faraday effect

In Fig. 5.2 the vectors $\vec{H}_+$ and $\vec{H}_-$ are represented at a fixed point in time for $z = 0$ and at a location $z = z_1$ farther away in the direction of propagation. The vectors $\vec{H}_+$ and $\vec{H}_-$ at location $z_1$ lag in their rotation behind those at the location $z = 0$ by the spatial angle $\varphi_+ = \beta_+ \cdot z_1$ or $\varphi_- = \beta_- \cdot z_1$. The direction of the total field intensity $\vec{H} = \vec{H}_+ + \vec{H}_-$ is rotated at location $z_1$ (in the case where we neglect the attenuation difference) by the angle

$$\vartheta = \frac{1}{2}(\varphi_+ - \varphi_-) = \frac{1}{2}(\beta_+ - \beta_-) \cdot z_1 \ . \tag{5.27}$$

In the positive and negative directions of rotation, circulating waves in ferrite exhibit different propagation constants. Subsequently, a linearly polarized wave can be split into two circularly polarized waves of equal amplitude and phase with opposite rotation directions, as is illustrated by Fig. 5.7. Next, if a linearly polarized wave encounters a premagnetized ferrite in its direction of propagation, then clockwise and counterclockwise circularly polarized partial waves with varying propagation constants propagate (circular double-refraction). In the case of premagnetization until saturation occurs, the ferrite exhibits the complex permeability $\mu_+$ with respect to the clockwise-polarized alternating field $H_+$ and the permeability $\mu_-$ with respect to the counterclockwise-polarized alternating field $H_-$, according to Eq. (5.26).

A linearly polarized wave, therefore, undergoes a rotation of the polarization plane according to Eq. (5.27) in the case of a ferrite moving in the direction of an installed static magnetic field. We call this ferromagnetically conditioned polarization plane rotation (use of Lord Rayleigh's idea by Hogan and Roberts, 1952, as a Faraday turner) as the Faraday effect, as in optics, according to the (much smaller) rotation of the polarization plane in glass. Discovered in light waves by Michael Faraday in 1846, whereupon a static magnetic field acts in the light's direction of propagation, it is based on the diamagnetic precession of the entire atomic electron cloud. The amount and direction of the wave polarization rotation in ferrite depend on $\mu_+$ and $\mu_-$ and, therefore, on the frequency and $H_0$. The direction of the Faraday rotation is different above and below the resonant frequency of the ferrite.

Equation (5.26) demonstrates that at an alternating field frequency $\omega = \omega_0$, the electron spin precession forced by the rotating alternating field $H_+$ in the clockwise direction (i.e., in the same direction as the free electron spin precession) enters resonance ($\omega_0 = \gamma \cdot H_0$ depending on the static magnetic field intensity $H_0$ of the free electron spin precession (L'Armor frequency). In Fig. 5.3 the real and imaginary parts of the relative permeabilities $\mu_\pm = \mu'_\pm - j\mu''_\pm$ are represented as a function of the constant field intensity $H_0$ for a fixed frequency $f = \omega/2\pi$. The effective permeability $\mu_+$ for the right-hand circularly polarized wave whose rotation direction agrees with the preferred (right-hand directed with respect to the static magnetic field) precession of the ferrite's electrons, passes through a location of resonance at $H_0 = \omega/\gamma$, which means at $\omega = \omega_0$, where the real part $\mu'_+$ changes sign and the imaginary component $\mu''_+$ becomes very large. In the resonant region, a strong interaction takes place between this wave and the ferrite. The path of the imaginary part of the permeability describes the attenuation behavior of the gyromagnetic material. The attenuation constant $\alpha$ increases sharply for the right-handed wave in the region of gyromagnetic resonance as a result of the removal of energy by the ferrite (resonance absorption). The precession movement of the electrons is strongly excited and the losses created by this as a result of the inner friction are counterbalanced by a corresponding absorption of the electromagnetic energy of the wave, whereby the ferrite material heats up. The alteration of the real part $\mu'_+$ causes a variation of the phase constant, the wavelength, or the phase velocity, in the region of negligible losses in the ferrite, corresponding to the relation which is then valid:

$$\beta_+ = 2\pi/\lambda_+ = \omega \cdot \sqrt{(\epsilon \cdot \mu_+)}$$

$H_{res}$ $\omega/\Gamma$ is the resonant field intensity for which we obtain a precession frequency $\omega_0$ which coincides with the alternating field frequency. Then we obtain the strongest excitation of the precession with the largest precession pin aperture angle and the largest losses. For the electron precession forced by the counterclockwise alternating field $H_-$, we obtain no resonance. The magnetic field rotates in the contrary sense (left-handed) of the preferred precession rotation direction of the ferrite and, therefore, a mutual influence hardly takes place. The permeability $\mu_-$ depends only very little on the premagnetization of the ferrite. Because $\omega_0 \gg 1/\tau = \omega\alpha$, usually $\tau \approx 10^{-8}$ s, the imaginary component remains $\mu''_- \approx 0$ and, therefore, the attenuation of the counterclockwise wave remains low and almost unaffected by the premagnetization of the ferrite. The curves $\mu'_+ + \mu''_+ = f(H_0)$ describe the frequency dependence of $\mu'_+$ and $\mu''_+$ at constant $H_0$ (at the point of resonance: $\omega = \omega_0$) in the case of a reversed abscissa direction.

In the case of low premagnetization below saturation, $\mu''_+$ and $\mu''_-$ increase (low field losses): the Weissian zones are no longer polarized in the same direction and the Bloch walls between these zones conduct three-dimensional oscillations with the alternating field frequency and, therefore, remove energy from it.

In addition to the permeability of a ferrite pellet, which is small compared to wavelength the demagnetization factors or the pellet for all three directions in space are equal to $4\pi/3$ in the homogeneous magnetic field, Eq. (5.26) is also valid for the case in which plane circularly polarized waves propagate in the $H_0$ direction in a ferrite medium of infinite length. In the case of a non-spherical ferrite probe of finite dimensions, the demagnetization fields must be considered. We obtain similar dependencies for the permeability numbers of $H_0$, as in Fig. 5.3, whereby $\omega_0$ now deviates from $\Gamma \cdot H_0$.

If, proceeding from Eq. (5.23), we investigate the wave propagation in a ferrite medium of infinite length for the case of propagation perpendicular to the direction of the premagnetization field intensity $\vec{H}_0$, then we obtain with $\vec{H}$ parallel to $\vec{H}_0$ the propagation constant

$$\gamma = j\omega \cdot \sqrt{\mu_0 \cdot \varepsilon} \,. \tag{5.28}$$

Since $\mu = \mu_0$, therefore, no gyromagnetic influence takes place upon the wave. With $H$ perpendicular to $H_0$ we obtain the expression

$$\gamma = j\omega \cdot \sqrt{\frac{\mu_1^2 - \mu_2^2}{\mu_1} \cdot \varepsilon} \tag{5.29}$$

with $\mu_1$, $\mu_2$ according to Eqs. (5.21) and (5.22). We designate such a medium, in which the propagation constant depends on the polarization direction of the wave, as double-refraction.

The direction-controlled behavior of the ferrite, which is expressed by the permeability curves according to Fig. 5.3 as non-reciprocal, is used in the construction of non-reciprocal components. A non-reciprocal component influences

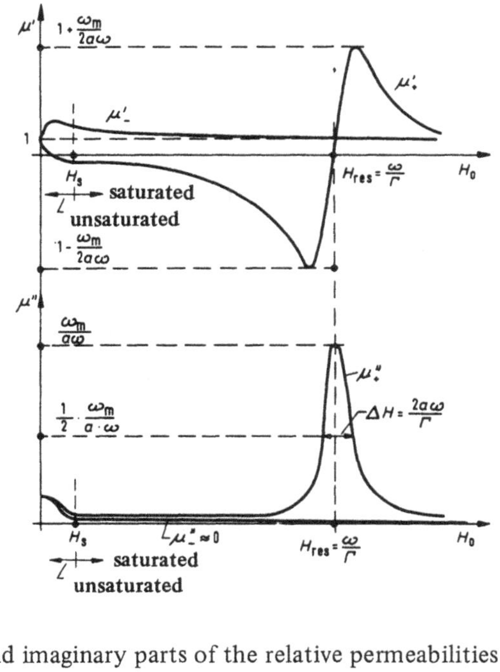


**Fig. 5.3** Real and imaginary parts of the relative permeabilities $\mu_\pm = \mu'_\pm - j\mu''_\pm$

a wave propagating in the forward direction (incident wave) in a different manner than it does a wave propagating in the backward direction (reflected wave). Here we take advantage of the strong difference between the imaginary parts $\mu''_+$ and $\mu''_-$ of the permeabilities in the resonant range $H_0 \approx \omega_0/\Gamma$, or the difference in magnitude of the real part $\mu'_+$ and $\mu'_-$ at $H_0 \ll \omega_0/\Gamma$. In the first instance, we obtain a different attenuation and, in the second, an effect of various phase velocities for right- and left-handed circularly polarized waves.

Some of these ferrite components will be illustrated in the following. They can be classified according to the fundamental principles of non-reciprocal resononce absorption, of the non-reciprocal Faraday rotation, of the non-reciprocal phase shift, or the non-reciprocal field suppression. The most important non-reciprocal components are the directional or one-way line, the directional fork or circulator, and the directional phase shifter. Direction lines which are non-reciprocal attenuation components can be realized with the help of gyromagnetic resonance absorption; the Faraday rotation, or the non-reciprocal field distortion and circulators in the use of the Faraday rotation; the non-reciprocal phase shift; and the non-reciprocal field distortion.

Components, for which the ferrite is premagnetized perpendicular to the wave's propagation direction are also called cross field components and those with a premagnetization parallel to the direction of propagation, as longitudinal field components.

## 5.2 NON-RECIPROCAL FERRITE COMPONENTS

### 5.2.1 Utilization of Resonance Absorption

We utilize the different attenuation behavior of the ferrite with regard to right- and left-handed circularly polarized waves with the resonance absorption at $H_0 = \omega/\Gamma$ in the resonance direction line according to Fig. 5.4.

A *directional* or *one-way line* (isolator) which causes virtually unattenuated transmission ($|\underline{S}_{21}| \approx 1$) in one transmission direction (forward direction) and displays a high attenuation ($|\underline{S}_{12}| \approx 0$) in the opposite direction (backward direction). Therefore, the S-matrix of an ideal one-way line, reflection-free at both gates ($S_{11} = S_{22} = 0$) (with a phase rotation 0 in the transmission direction), reads

$$\underline{S} = \begin{vmatrix} 0 & 0 \\ 1 & 0 \end{vmatrix} . \tag{5.30}$$

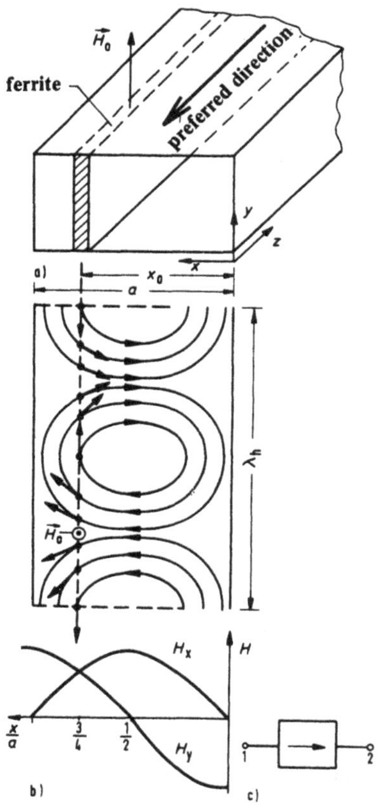


**Fig. 5.4** (a) Resonance directional line; (b) Circular polarization of the $H_{10}$-wave at $x_0$; (c) Radio-circuit symbol of a directional line

In a practical sense, this idealized behavior is not possible since $|\underline{S}_{12}|^2 + |\underline{S}_{22}|^2 = 1$ and $\underline{S}_{11}\underline{S}_{12}^* + \underline{S}_{21}^*\underline{S}_{22}$ must equal 0 [see Eqs. (2.47) to (2.49)].

The measurement of the transmission attenuation $a_D$ and the stop-band attenuation as

$$a_{D,s} = 10 \log \frac{P_{ein}}{P_{aus}} \text{ dB} \tag{5.31}$$

takes place at an adapted input and output.

The directional line (symbol of radio-circuit element, Fig. 5.4c) is used, for example, in the decoupling of generator and load or of the stages of an amplifier.

The necessary magnetic rotating field of the resonance directional line can be obtained for a rectangular hollow waveguide with $H_{10}$-wave propagation laterally in a plane parallel to the narrow side, as Fig. 5.4b illustrates. We recognize that for a wave propagating in the $+z$-direction at a fixed point for $x > a/2$, we obtain a positive rotation of the magnetic field intensity vector in the $xz$-plane with reference to the $H_0$-direction; for a wave propagating in the opposite direction, the $H$-vector rotates in the negative direction. For the relationship between the cross field component $\underline{H}_x$ and the longitudinal field component $\underline{H}_z$, we obtain according to Eqs. (1.181) and (1.182):

$$\frac{\underline{H}_x}{\underline{H}_z} = \pm \mathrm{j} \frac{2a}{\lambda_h} \cdot \tan \frac{\pi x}{a} . \tag{5.32}$$

Purely circular polarization with $\underline{H}_x = \pm \mathrm{j}\underline{H}_z$ is obtained according to Eq. (5.32) at

$$x_0 = \frac{a}{\pi} \cdot \operatorname{arc\,tan} \frac{\lambda_h}{2a} . \tag{5.33}$$

The magnetic field is circularly polarized at approximately 1/4 of the waveguide broad side in the planes parallel to the broad side. For $x \neq x_0$ the field is elliptically polarized.

The different rotation direction of the $H$-field for both directions of wave propagation yields a non-reciprocal (i.e., direction-controlled) attenuation in a thin ferrite plate premagnetized with $H_0$, which is introduced at $x_0$ in the waveguide. The rotating field is somewhat elliptically deformed by the demagnetized field of the ferrite plate. This can be approximately compensated by a small displacement of the ferrite strip with respect to $x_0$, according to Eq. (5.33), and we then obtain an almost circular magnetic field in the plate. The static magnetic field perpendicular to the alternating field $H$ can be produced with a permanent magnet. $H_0$ should be selected such that when operating the one-way line the resonance of the electron spin precession in the ferrite can be adjusted.

A wave propagating in the backward direction of the resonance absorption line ($+z$-direction in Fig. 5.4) has a positive circular magnetic field (clockwise at a fixed location with reference to $\vec{H}_0$) in the ferrite, whose rotation direction agrees with that of the electron spin precession. At resonance, a strong interaction takes place between the wave magnetic field and the attenuated precession of the magnetic momentum with a strong excitation of the positive electron spin precession. Here the energy of the wave is almost entirely absorbed in the ferrite and transformed into heat. With a wave propagating in the opposite direction, the direction of the polarization rotation changes with respect to the premagnetization direction. The sense of rotation of the negative circular polarization obtained at the location of the ferrite is opposite to the preferred precession movement and no excitation of this movement occurs; the wave only excites the weak, almost non-absorbing negative spin precession and is, therefore, barely attenuated.

This directional action can be increased if we attach a second ferrite strip symmetrically to the middle of the waveguide, which must be premagnetized in the opposite direction because of the opposite circular polarization at this location (Fig. 5.5).

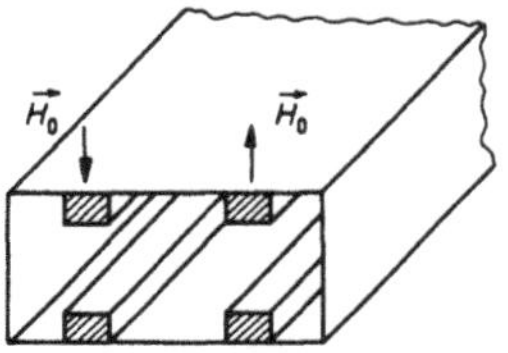


**Fig. 5.5** Resonance directional line with two-sided arrangement of ferrite strips

The heat created in the ferrite plates of a resonance directional line by the absorption of microwave power can be easily eliminated over the adjoining waveguide wall and such a directional line can be constructed for large loads.

In order to attain non-reciprocal transmission characteristics with ferrites in coaxial transmission lines, the magnetic field of the coaxial wave, which does not show locations of circular polarization, must be distorted so that it acquires a component in the direction of propagation. A partial excitation of an *H*-wave with regions of circular polarization is brought about. This takes place through a partial (e.g., half-sided) filling of the line with a dielectric material, which can also be the ferrite itself.

The slot line and the coplanar line (double-slot line) are suitable for the construction of non-reciprocal ferrite components, whose magnetic field shows a component in the direction of propagation (Fig. 1.26f), whereby we obtain regions of elliptical polarization.

The backward attenuation of resonance absorption lines is usually approximately 20–40 dB and the forward attenuation $\geqslant 1$ dB.

One-way lines are used for decoupling devices. Therefore, in the case of a poorly adapted load, in order to keep the wave away from the output, from a generator for example; in order to avoid frequency pulling by a charge; in decoupling of amplifier stages in order to insure stable amplification; or, generally, to improve the matching of components.

In addition to the principle of resonance absorption, we also use the Faraday effect and the field suppression effect in constructing directional lines.

### 5.2.2 Faraday Turners

For the partial waves with positive and negative circular polarization of a linearly polarized wave in an unlimited, low-attenuation ferromagnetic medium, the following is valid

$$\gamma_{\pm} = j\omega \cdot \sqrt{\varepsilon \cdot \mu_{\pm}} \approx j\beta_{\pm} \ . \tag{5.34}$$

Because of the different phase rotation $\beta$ of the two opposite-rotating circularly polarized waves, the Faraday effect occurs with a rotation of the polarization plane, as represented in Fig. 5.2. We use this effect in the construction of polarization rotators or non-reciprocal phase shifters by choosing the magnetic constant field intensity $H_0$ so far below the resonant field intensity $\omega_0/\Gamma$, that the difference between the real components $\mu'_+$ and $\mu'_-$ of the ferrite permeabilities in the case of a clockwise or counterclockwise magnetic field becomes as large as possible with $\mu'_- > \mu'_+ \approx 0$, and with $\mu''_+ \approx 0$. The absorption losses remain low (compare with Fig. 5.3). Figure 5.6 shows an arrangement, based on this principle, of the polarization plane rotation, a Faraday *turner* or *rotator,* which operates with $H_{11}$-wave propagation in a circular hollow waveguide. A

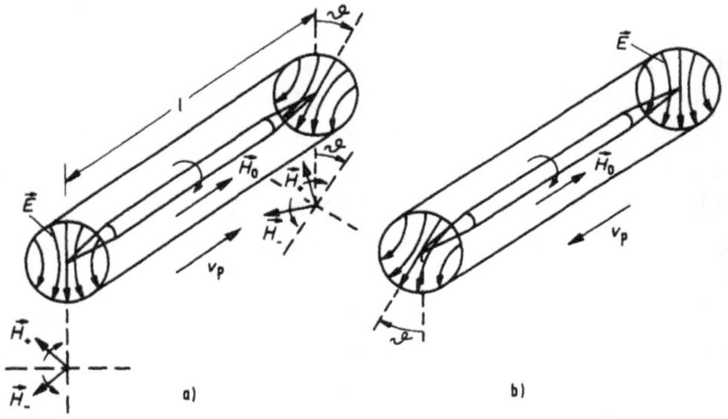


**Fig. 5.6** Faraday turners: (a) feeding in the front, (b) feeding in the back

cylindrical ferrite bar is located in the waveguide axis whose ends are sharpened in order to reduce reflection. With a spool surrounding the waveguide, the ferrite bar is premagnetized in the axis direction with the magnetic constant field $H_0$ corresponding to the specified criterion.

The ferrite bar in the middle of the waveguide is exposed to the field of the $H_{11}$-wave, which is linearly polarized there, and which can be considered plane. This linearly polarized wave can be split with the transverse magnetic field $\vec{H}$, as Fig. 5.7 illustrates, into two opposite-rotating circular waves with the transverse magnetic fields $H_+$ and $H_-$. The positive rotating wave propagates in the

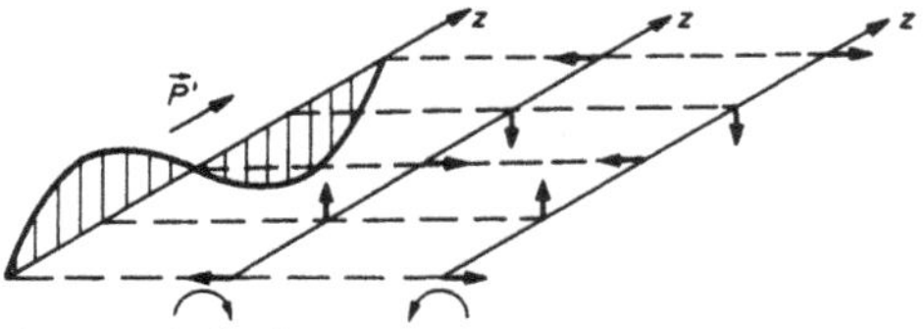


linear polarized wave =
clockwise + counterclockwise
circularly polarized wave

**Fig. 5.7** Dissection of a linear polarized wave into two opposite-rotating circularly polarized waves

ferrite corresponding to the decreased permeability number $0 < \mu'_+ < 1$ with the decreased [Eq. (5.34)] phase constant $\beta_+ < \beta$, and for the circular partial wave with a negative rotation direction, an increased phase constant $\beta_- > \beta$ is valid because of the increased permeability number $\mu'_- > 1$. The positive circularly polarized wave, therefore, undergoes a three-dimensional rotation of $\varphi_+ = \beta_+ \cdot l$ along the length $l$ in the region of the ferrite bar, and the negative circular wave, because of $\beta_- > \beta_+$ larger field rotation, by $\varphi_- = \beta_- \cdot l$. Therefore, the phase difference of both waves according to the length $l$ is $\varphi = \varphi_- - \varphi_+ = (\beta_- - \beta_+) \cdot l$. If we bring the two circularly polarized waves back together into the resulting linearly polarized wave with $\vec{H} = H_+ + H_-$, then it is rotated at the end of the ferrite cylinder in its polarization plane with respect to the polarization plane at the input at the three-dimensional angle

$$\vartheta = \frac{\varphi}{2} = \frac{1}{2}(\beta_- - \beta_+) \cdot l \tag{5.35}$$

in the $H_0$-direction as seen from the right (at $f > f_{res}$, from the left at $f < f_{res}$); because of the equal attenuation of the circular polarized waves with $\mu''_+ \approx \mu''_- \approx 0$ the resulting wave maintains a linear polarization. In this connection, the direction of the polarization plane rotation does not depend on the wave propagation direction with reference to the $\vec{H}_0$ direction, but rather is a function of the static magnetic field $H_0$ which determines the direction of the electron spin precession. Therefore, a wave propagating in the opposite direction of the $H_0$ direction produces the same excitation of the electron spin precession in the ferrite with a polarization plane also rotated to the right with respect to the $\vec{H}_0$-vector. With reference to the wave propagation direction, however, this signifies a rotation to the left. The component exhibits, therefore, a non-reciprocal, or direction-controlled, transmission. The magnitude of the rotation can be varied with the intensity of the static magnetic field.

We usually allow the circular waveguide cross section at the ends of the rotator to pass over into a rectangular cross section in order to attach the component directly onto the customary $H_{10}$ rectangular hollow waveguide.

Because of the central arrangement of the ferrite in the waveguide, the heat created in the ferrite cannot be effectively eliminated. Therefore, the component is only suitable for relatively small charges (note the temperature variation of the ferrite!)

With the help of the Faraday turner described above, a hollow waveguide gyrator can be constructed. A *gyrator* is a non-reciprocal dual-gate for which the transmission angles of the two transmission paths differ by $\pi$. To this end, we choose a Faraday rotation of 90° and rotate the polarization by an additional 90° with a waveguide component that is twisted together at 90°, and connecting to the turner so that for this transmission direction a total rotation of 180° takes place; therefore, a phase reversal. For the opposite transmission path, the mechanical rotation of the polarization plane in the twisted waveguide component is cancelled by the Faraday rotation which, of course, maintains its rotation direction with respect to the $\vec{H}_0$ direction. Therefore, we obtain for the two transmission directions a phase constant difference of $\pi$.

*Faraday Isolator*

The principle of the Faraday rotation can also be applied in the construction of direction lines. Figure 5.8 shows a one-way line (isolator) constructed with the use of a 45° Faraday turner. The $H_{10}$-wave fed into input gate 1 is transformed into an $H_{11}$-wave by a transition from a rectangular to a circular waveguide cross section. The connecting Faraday turner rotates the polarization direction of the $H_{11}$-wave by 45° in the clockwise direction (at $f > f_{res}$, sense of rotation to the

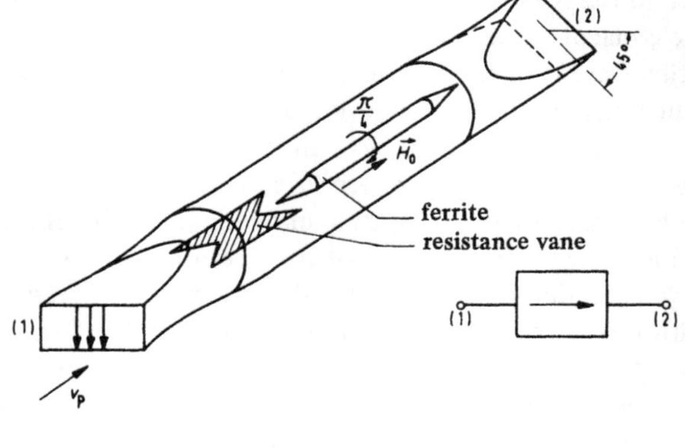


**Fig. 5.8** Faraday directional line and radio-circuit symbol

left at $f < f_{res}$). Therefore, the wave can enter the rectangular waveguide connected to a second transition and rotated at 45°. With a twisted waveguide component, the $H_{10}$-wave at the output of the isolator can be brought back into the same polarization direction as at the input. The resistance vane mounted transversely in the turner input does not have an effect on the wave approaching gate 2, since its electrical lines of force run perpendicular to the attenuation layer. A wave fed in gate 2 (reflected wave) is rotated to the left by 45° in its polarization with a Faraday turner with respect to the direction of propagation. Its polarization direction then runs parallel to the resistance vane and to the waveguide broad side of gate 1. The wave is strongly attenuated in the resistance vane, since its electrical field runs parallel to the vane and, through the convection currents created in it, the energy of the wave is transformed into heat. The wave cannot enter the input of the rectangular waveguide because of the unsuitable polarization, and is, therefore, reflected at the input and weakened again at the absorption layer while returning.

The component, therefore, exhibits a non-reciprocal transmission behavior with a low transmission attenuation in one transmission direction (gate 1 → gate 2) and a high stop-band attenuation in the opposite direction. Since with the thin resistance vane the absorbed power cannot be effectively eliminated, this isolator is only employed for small loads.

*Faraday Circulator*

The arrangement according to Fig. 5.8 can be extended by the coupling of two additional rectangular hollow waveguides into a four-port circulator (Fig. 5.9).

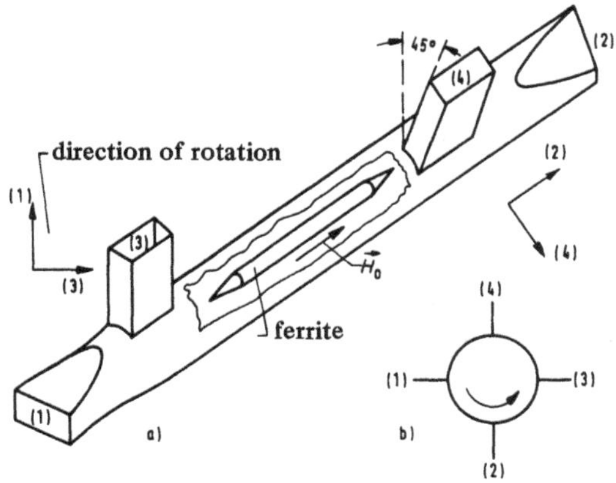


**Fig. 5.9** (a) 4-port Faraday circulator; (b) Radio-circuit symbol

A *circulator (directional fork)* is a non-reciprocal component with three or more gates which allows transmission in each case to only one gate in a given sequence, whereas no charge is tightly coupled to the other gates. Therefore, in the case of an ideal triple-gate or three-port circulator, whose radio-circuit symbol is represented in Fig. 5.15c, with the sense of rotation chosen there (circulation direction), $1 \rightarrow 2 \rightarrow 3$, a wave fed in gate 1 at a non-reflecting termination of gates 2 and 3 is only transmitted to gate 2, and gate 3 is "insulated." This characteristic is maintained in the case of a cylindrical transposition of the gates: if gate 2 is fed and gates 1 and 3 are adapted, then it is transmitted to gate 3, and gate 1 is decoupled; a corresponding procedure is valid for the transmission path $3 \rightarrow 1$.

Therefore, the S-matrix of this ideal triple-gate circulator with lossless transmission (transmission attenuation 0), for which no transmission takes place in the opposite directions (stop-band attenuation $\rightarrow \infty$) and which is adapted at all gates, reads

$$\underline{S} = \begin{pmatrix} 0 & 0 & e^{j\varphi} \\ e^{j\varphi} & 0 & 0 \\ 0 & e^{j\varphi} & 0 \end{pmatrix} . \tag{5.36}$$

In this connection, the reflectivities are $S_{11} = S_{22} = S_{33} = 0$, the operating transmission factors are $S_{12} = S_{23} = S_{31} = 0$, and the operating transmission factors in the direction of transmission $\underline{S}_{21} = \underline{S}_{32} = \underline{S}_{31}$ equal 1 and the impedance angle $\varphi$ (at a suitable distance of the gates from the center $\varphi = n \cdot 2\pi$, $e^{j\varphi} = 1$).

The measurement of the transmission and stop-band attenuations for the various transmission paths takes place in each case with a matched termination of all gates. In practice, the transmission attenuation is usually $\leqslant 1$ dB and the stop-band attenuation $\geqslant 20$ dB.

In Fig. 5.9, a four-gate Faraday circulator is represented. It operates with a 45° turner and a polarization deflector at both the input and the output. The polarization turner is comprised of a circular waveguide section with $H_{11}$-wave propagation, in which a premagnetized ferrite is axially mounted. The ferrite is premagnetized with the help of an outer magnetic field such that the polarization direction of the incoming waves propagating through the ferrite-loaded waveguide section passes over to both sides in the rectangular hollow waveguides, which are rotated by 45° with respect to each other. At the ends of the circular waveguide section, an additional gate (3 and 4) is attached in each case through the rectangular waveguide arms, which are respectively arranged perpendicular to the neighboring gate (1 and 2). The rectangular waveguide terminal cross sections are proportioned such that at the operating frequency only an $H_{10}$-wave is able to propagate. The rectangular waveguide 3 is decoupled from waveguide 3 by the polarization directions differing by 90°; the same is valid for the hollow waveguide 2 and 4. In the case of the premagnetization direction chosen in Fig. 5.9 and the respective geometry, a wave fed in gate 1, which is rotated by 45° to the right ($f > f_{res}$, $\beta_+ \cdot l > \beta_- \cdot l$) in its polarization direction in the ferrite zone, can only be transmitted to gate 2. Only this arm can absorb its polarization. Since the sense of the polarization plane rotation is fixed with reference to the premagnetization direction and independent of the wave's propagation direction, a wave directed to gate 2 will also be rotated to the right by 45° with respect to the $\vec{H}_0$ direction, and arm 3 has conditions suitable for absorption of its polarization. Thus, it can be decoupled there. Corresponding considerations are valid for the transmission paths 3 → 4 and 4 → 1. The rectangular waveguide 1 is uncoupled from waveguide 3 by the polarization direction differing by 90°; this also holds true for hollow waveguides 2 and 4. A low-reflection coupling of the lateral arms to the circular hollow waveguide in a wide frequency range can be obtained with the help of matching elements.

If gates 3 and 4 are closed reflection-free, then a wave can propagate from gate 1 to gate 2. In the opposite direction, however, high attenuation occurs; the arrangement functions in this way as a directional line (isolator).

According to the principle of the Faraday circulator, a variable *ferrite fader* can be constructed. If we vary the polarization plane rotation with the help of the attached magnetic constant field, then the charge transmitted from gate 1 to gate 2 changes. The fader is usually constructed such that gates 1 and 2 are not rotated with respect to each other, and therefore without a rotation of the polarization plane no attenuation occurs. Arms 1 and 3 are closed reflection-free, so that the undesired charge is absorbed there.

### 5.2.3 Non-Reciprocal Phase Shift

The arrangement of a rectangular waveguide with asymmetrically situated ferrite strips (Fig. 5.4a), or two-sided ferrite strips with opposite premagnetization direction (Fig. 5.5), can be used as a phase shifter if operated in the low-loss range ($\mu_+'' \approx 0$) with a static magnetic field corresponding to $H_0$ well below the resonant field intensity (low-field operation ($\omega > \omega_0$) or above (high-field operation ($\omega < \omega_0$)). Transmission is now possible in both directions, whereby we obtain, for both transmission directions, different phase shifts $\beta_+ \cdot l$ [for propagation in the $z$-direction in Fig. 5.4) or $\beta_- \cdot l$ (propagation in the $-z$-direction, ($l$ = length of the premagnetized ferrite strips)] as a result of the difference between $\mu_+'$ and $\mu_-'$. This arrangement operates as a non-reciprocal phase shifter. In the low-field operation the premagnetization field intensity must remain so high that the ferrite is saturated. The phase shift can be varied by changing the static magnetic field.

The S-matrix of an ideal (that means, non-dissipative and adapted non-reciprocal) phase shifter with zero phase shift in the backward transmission direction, $2 \rightarrow 1$, reads

$$\underline{S} = \begin{pmatrix} 0 & 1 \\ e^{-j\Delta\varphi} & 0 \end{pmatrix} . \tag{5.37}$$

The forward phase shift $\Delta\varphi$ depends on the intensity of the premagnetization field. The radio-circuit symbol of a non-reciprocal phase shifter is represented in Fig. 5.10c.

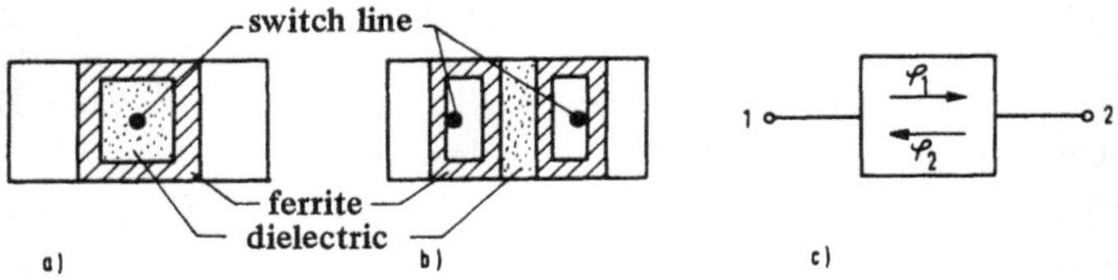


**Fig. 5.10** Non-reciprocal phase shifters: (a) simple toroid, (b) double toroid, (c) radio-circuit symbol

Figure 5.10 shows two type models of non-reciprocal phase shifters with ferrite toroids in a rectangular hollow waveguide (a. simple toroid, b. double toroid). By using the ferrite toroid in a microstrip or coaxial transmission line, the necessary magnetic longitudinal component is obtained by an asymmetrical loading of the line with the ferrite or dielectric material.

One possibility in the construction of gyromagnetic components in microstrip applications is also obtained with an irregularly shaped conductor on a ferrite substrate, whereby we obtain a 90° shift of the currents at these locations and, therefore, a circular polarization (*H*-fields displaced temporally and spatially by 90°) by a length of $\lambda_L/4$ of the two line pieces connecting the two conductor paths lying across from each other.

With suitable sizing, we obtain a difference in the transmission angle of $\pi$ (phase reversal) for both transmission directions. This component is called a *gyrator.* This kind of gyrator is often used as a circulator in connection with 3-dB directional couplers.

Figure 5.11 shows the diagram of a *circulator* which is comprised of a 3-dB rectangular waveguide directional coupler, two non-reciprocal phase shifters, and a reciprocal 90° phase shifter (dielectric $\lambda/2$ plate). The coupling between the primary and the secondary lines of the directional coupler takes place through

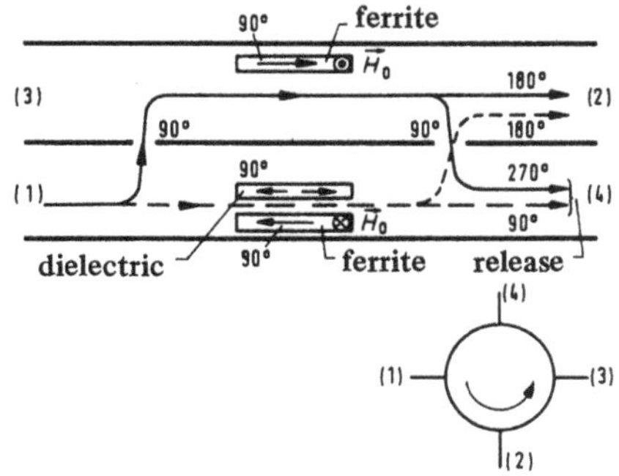


**Fig. 5.11** 4-port circulator using a directional coupler and phase shifters

two coupling holes in the common narrow side of both rectangular hollow waveguides. The ferrites are premagnetized in opposite directions perpendicular to the conducting plane. In Fig. 5.11, the phase relationships are outlined for the case of feeding in gate 1, which leads to a transmission toward gate 2. The 90° phase change is to be considered for the case of the wave passing through the coupling apertures of the directional coupler. On the whole, we obtain the transmission diagram specified in the radio-circuit symbol.

In another variation, a non-reciprocal phase shifter is constructed only in the secondary line, which causes an additional phase shift of 180° in one direction and no phase shift (gyrator behavior) in the other direction, and does so also as a laterally dislocated ferrite bar. As designated in Fig. 5.11, and by taking into consideration an additional phase shift along the right-hand transmission path through the ferrite, we obtain a transmission diagram with the sequence 1 → 4 → 3 → 2.

The bandwidth of the circulators described can be increased by extending the two coupling locations into multiple-hole couplings, as is described for the case of multiple-hole couplers (see Ch. 3).

### 5.2.4 Utilization of Field Distortion (Field Suppression)

For the undisturbed case of the rectangular waveguide, the electrical field intensity of an $H_{10}$-wave is distributed in the form of a sine semicircle over the waveguide broad side (Fig. 1.17a). With a premagnetized ferrite plate mounted on the outside of the waveguide middle (Fig. 5.12), we learn that for the incident wave a relatively small electrical field intensity is adjusted parallel to the plate, and for the reflected wave, on the other hand, a high field intensity. We

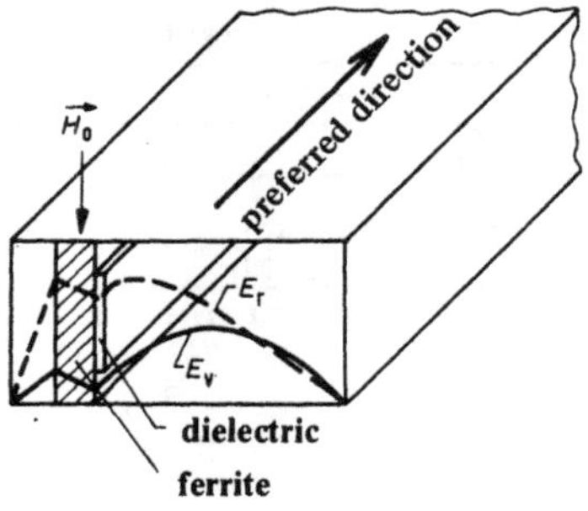


**Fig. 5.12** Field-suppression line

may then attach an attenuation layer parallel to the ferrite plate, so that the reflected wave is essentially more sharply attenuated than the incident wave, and the arrangement operates as a *one-way line.*

The premagnetization is chosen such that the real part of the permeability for the positive circularly polarized wave is $\mu'_+ \leqslant 0$, so that the ferrite has a reflecting effect for the wave; whereas for the negative circular waves $\mu'_-$ is positive and wave propagation is hampered in the ferrite. The determinant value $\mu''_+$ for the losses in the ferrite remains small.

Figure 5.12 shows a one-way line which is based on the field distortion. At the location of the circular polarization of the $H_{10}$-wave of a rectangular hollow waveguide, a ferrite strip premagnetized transverse to the propagation direction is attached. The rotation direction of the circular polarization depends on the propagation direction of the waveguide wave. In one propagation direction a larger portion of the energy flux is concentrated in the ferrite, whereas in the opposite propagation direction very little energy is transmitted in the ferrite. With a resistance layer mounted on the ferrite strip in a plane, in which the relationship between the electrical field intensities for the backward and forward directions is as large as possible, we obtain a correspondingly high stop-band attenuation, or respectively, for the opposite transmission direction, a low transmission attenuation. The ferrite, therefore, does not serve the attenuation of the wave in the high-resistance direction as in the case of the resonance-absorption directional line, but rather serves a field concentration in the ferrite so that the wave can be absorbed in the resistance strip. The arrangement operates with a wider bandwidth than the resonant one-way line.

Figure 5.13 shows a four-port *circulator,* based on the field suppression, and which works with a directional coupler. The coupling between the primary and the secondary line takes place over the longitudinal high-frequency magnetic field. A wave fed in arm 1 is completely coupled in the secondary line and leaves by gate 2. On the other hand, in the case of a wave directed to gate 4, the longi-

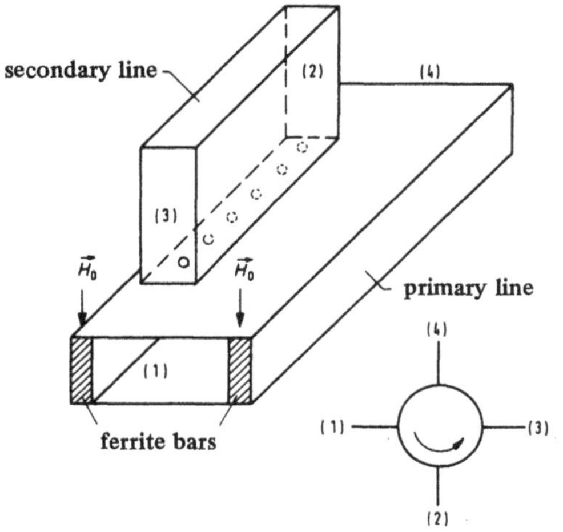


**Fig. 5.13** Field-suppression circulator

tudinal magnetic field is zero at the coupling location and the wave is transmitted to gate 1. On the whole, we obtain the transmission diagram specified in the radio-circuit symbol.

A similarly constructed three-port rectangular waveguide-circulator functions according to this principle as well, for which the transfer substation is also laterally dislocated in the broad side of the primary line and is formed as a longitudinal slot. From there, a rectangular waveguide side arm is coupled, which connects to the slot and is arranged perpendicular to the primary line. For the sense of direction, we obtain by analogy of Fig. 5.13 $1 \rightarrow 4 \rightarrow 3$.

A lower frequency limit (approximately 1 GHz) for use of non-reciprocal ferrite characteristics is given by the low static magnetic fields corresponding to the relation $H_0 = \omega_0/\Gamma$ for smaller precession resonant frequencies. These, then, do not suffice for saturation of the ferrite and it assumes reciprocal behavior. The upper frequency boundary (above 100 GHz) is essentially limited by the difficulties in producing the necessarily high magnetic inductions.

## 5.3 RECIPROCAL FERRITE COMPONENTS

Figure 5.14 shows some symmetrical arrangements of ferrites in the rectangular hollow waveguide with premagnetization perpendicular to the propagation direction. Because of the symmetrical arrangement of the ferrite strips and the

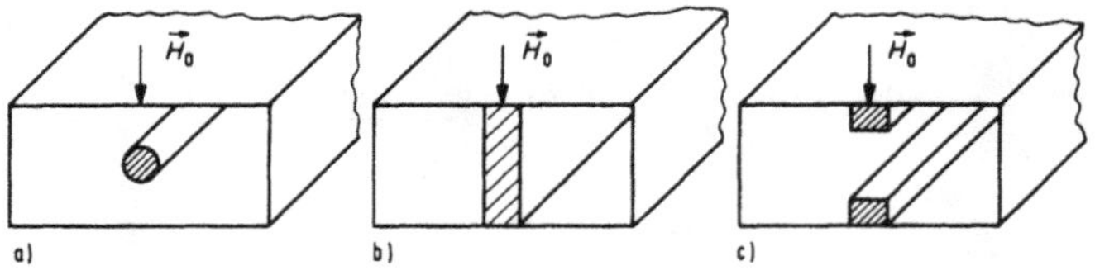


**Fig. 5.14** Reciprocal ferrite components (attenuator pads or phase shifters)

same direction of premagnetization with respect to both propagation directions; we obtain the same transmission behavior for both transmission directions; the dual-gates, therefore, display a reciprocal transmission behavior. For operating in the resonant range, we obtain reciprocal attenuation behavior because of the sharp increase of the imaginary part $\mu_+''$ of the active complex permeability (Fig. 5.3), and at frequencies well above (also occasionally below) the resonant frequency ($H \ll H_{res}$) in the low attenuation region (where imaginary part of $\mu$ is small), we obtain a reciprocal phase shift displacement corresponding to the real part $\mu_+$.

The ferrite strips of the examples in Fig. 5.13b, c are also mounted two-sided symmetrically to the waveguide middle (compare with Fig. 5.5) with the same premagnetization directions. Similar ferrite arrangements are also used in circular hollow waveguides and coaxial transmission lines. The arrangement corresponding to Fig. 5.14a, with a longitudinally magnetized ferrite bar in the longitudinal axis of an $H_{10}$ rectangular hollow waveguide, is known as a *Reggia-Spencer* phase shifter. Disadvantageous is the strong frequency dependence of its phase shift.

The S-matrix of an ideal (non-dissipative, reciprocal) phase shifter, with inherent reflectivity zero, reads

$$\underline{S} = \begin{pmatrix} 0 & e^{-j\Delta\varphi} \\ e^{-j\Delta\varphi} & 0 \end{pmatrix} \tag{5.38}$$

with the same phase shift $\Delta\varphi$ for both transmission directions, depending upon the size of the premagnetization field.

By variation of the externally installed magnetic field, with the help of the current of a magnetic coil, the attenuation of the reciprocal attenuator pads can be changed by altering the imaginary part of $\mu$. Its maximum value is at $H_0 = H_{res}$. The reciprocal phase shift can be varied in the same manner by changing the real part of $\mu$. Such variable attenuator pads can be used, among other things, as interrupters or modulators for amplitude modulation, and variable phase shifters can be used as phase or frequency modulators.

## 5.4 JUNCTION CIRCULATORS

A junction circulator is comprised of a line bypass whose arms all lie in one plane and in whose center a ferrite is mounted, which is premagnetized perpendicular to the plane of the bypass. We find designs in coaxial transmission line, hollow waveguide, or stripline technology with three or four gates. The circulator is most often used in microstrip design because of its simple, compact construction and its high bandwidth, and it is used with three gates.

In the case of a *triple-gate* or *Y-circulator* (Y-junction circulator, according to H.N. Chait and T.R. Curry, 1958; explanation of the mode of operation by C.E. Fay, H.L. Comstock, H. Bosma), a cylindrical ferrite body is located in the center of the junction point formed as a resonator of three waveguides displaced three-dimensionally by 120° with respect to one another. The ferrite is premagnetized in the axis direction perpendicular to the plane of the waveguide with a magnetic constant field $H_0$ (usually of a permanent magnet). The ferrite operates in the low-loss region with $H_0$ above $H_{res}$ or $H_0 < H_{res}$, or below the range of the low-field losses. Usually, we work beneath $H_{res}$ since such high working frequencies can be attained.

Figure 5.15a shows the diagram of a Y-circulator in microstrip application. The circulator disk where the stripline displaced with respect to one another by

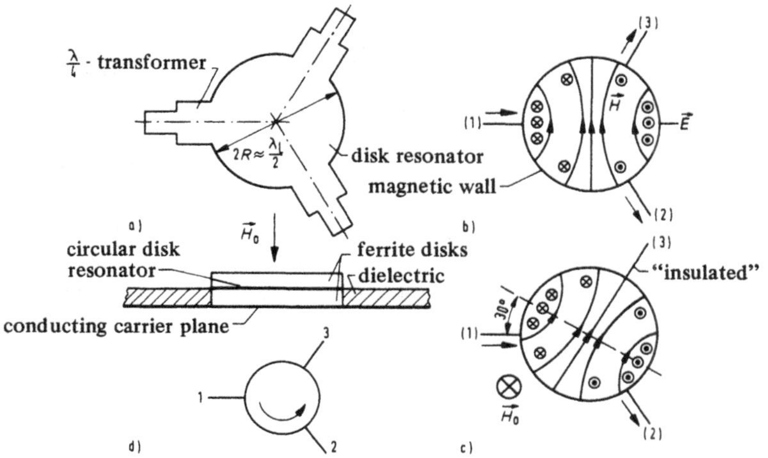


**Fig. 5.15** (a) Y-circulator in microstrip execution; (b) Field distribution at $H_0 = 0$; (c) Field distribution at premagnetization; (d) Radio-circuit symbol

120° converge, can be considered as a disk resonator. The lowest resonant frequency of the circular disk structure is obtained for the formation of the $E_{110}$-mode (dipole mode, $n = 1$). In this mode the electrical lines of force run perpendicular and the lines of the magnetic HF-field parallel to the disk plane. Figure 5.15b first shows the path of the electrical and magnetic lines of force (phase delay of $T/4$) of the standing wave excited by feeding in gate 1 in the event that the ferrite is not premagnetized. The formation of the standing wave (pure standing wave if the circulator is assumed to be non-dissipative) can be understood as the superposition of two waves of equal amplitude rotating oppositely in the circumferential direction, into which the wave fed can be separated. We obtain the field pattern as a superposition of the two equal opposite field configurations. With both partial fields, the magnetic *HF*-field in the middle of the disk is circularly polarized, with an increasing elliptical distance, and linearly polarized at the edge.

If the ferrite is premagnetized in the direction of the disk axis with a constant field, then different permeabilities $\mu_+$ and $\mu_-$ become active for the two opposite fields, and they are no longer resonant at the same frequency; hence, the modes separate. Whereas the two opposite modes degenerate without premagnetization – that is to say, they display the same resonant frequency – they are not degenerated by the effect of a premagnetization field in the direction of the symmetry axis of the arrangement. The resulting proper frequencies $f_+$ and $f_-$ depend on the intensity of the externally installed magnetic constant field. We now obtain a field circulation with a frequency which lies between the resonant frequencies of the non-degenerated modes. The mode with the field that rotates in the same direction as the electron spin precession ($E^+_{110}$-mode, short + mode) has the higher resonant frequency $f_+$. In the case of an excitation with the frequency which lies between the resonant frequencies of the two partial waves, the wave impedance of the + mode exhibits an inductive reactive component and that of the – mode (opposite sense of rotation with respect to the electron spin precession) with the lower resonant frequency $f_-$, a capacitative reactive component. At the operating frequency $f = (f_+ + f_-)/2$ of the circulator selected between the two resonant center frequencies, the capacitative and inductive reactances become equally large, and the total impedance becomes real. The two opposite partial waves are then equally strongly excited so that a standing wave is once again created as a superposition.

If we adjust the premagnetization and, therefore, the size of the resonant frequency division such that the impedance angle of both partial impedances $\phi = \arctan B/G$ at the operating frequency is 30°, then the field pattern of the standing wave is rotated by 30° with respect to the case without premagnetization, and is done so in the rotation direction of the + mode (Fig. 5.15c). For the + mode with its inductive reactive component, the maximum electrical field intensity is temporally and spatially displaced by 30° (since 1 rotation per period) with respect to the maximum $H$ at the input gate; for the - mode, we obtain a corresponding phase delay of $E$. Therefore, the maximum $E$ of both modes coincide at an angle of 30° with respect to the input gate, and we thus obtain a 30° rotation of the resulting field pattern in the + mode rotation direction. In this manner, the electrical field intensity at gate 3 becomes zero. At gate 3 (in the case of matching at gate 2) no wave is decoupled, it is "insulated"; gate 1 and gate 3 are decoupled; in practice, the attenuation is usually $\geqslant$ 20 dB. The arrangement operates as a resonator which transmits the power fed in gate 1 over its field toward gate 2. On the whole, we thus obtain transmission possibilities in the directions 1 → 2, 2 → 3, and 3 → 1 (for $H_0 < H_{res}$ and at $H_0 > H_{res}$ in the opposite direction of circulation; direction of rotation also dependent on the $H_0$ direction). The ferrite is operated below $H_{res}$ in the low-loss area.

For the S-matrix of the ideal triple-gate circulator, Eq. (5.36) is valid.

The center frequency of the circulator is

$$f_r = \frac{1}{2\pi} \cdot \frac{1{,}84}{R \cdot \sqrt{\varepsilon \cdot \mu_{eff}}} \, .$$

The low input impedance matching of the resonator is often undertaken with a $\lambda/4$ transformation line (Fig. 5.15a). Instead of using the ferrite disks intended in Fig. 5.15, the circuit is also constructed by using a ferrite substrate as column support.

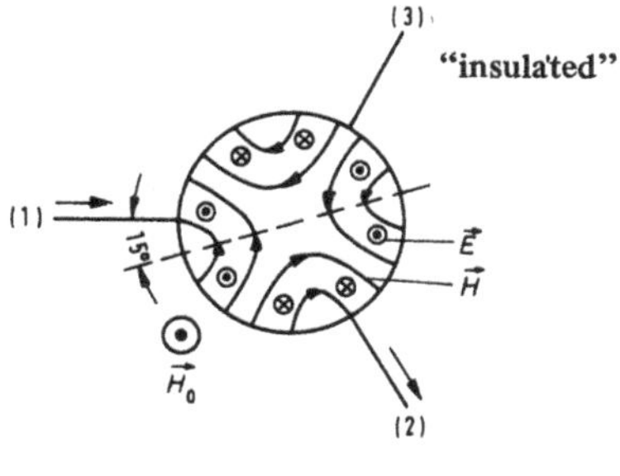


**Fig. 5.16** Y-circulator with disk resonator in the $E_{120}$-mode

Figure 5.16 shows the field pattern of a stripline circulator disk resonator of the $E_{210}$ disk resonance ($n$ = 2). Since two periods are lost on the circumference of the disk in this mode, the necessary three-dimensional rotation of the field here is 15° (30° electrically). The direction of the rotation in this mode agrees with the direction of circulation.

For *Y-circulators in rectangular waveguide applications*, we usually use a by-pass in the $H$-plane (plane of the broad sides). Figure 5.17 shows such an arrangement with the field pattern of an $E_{110}$ resonance, excited by an $H_{10}$-wave fed into gate 1. A cylindrical ferrite bar is located in the center of the bypass (cross section occasionally triangular in order to produce a larger bandwidth), which is premagnetized in the direction of the cylinder axis. The magnetic *HF*-field at the location of the ferrite is essentially circularly polarized. The operation of a circulator occurs as it does in the case of the stripline circulator. In order to increase the bandwidth, the bypass is often constructed with a reduced waveguide height (lower oscillation resistance) and adapted in a step-by-step transition to the normal waveguide profile.

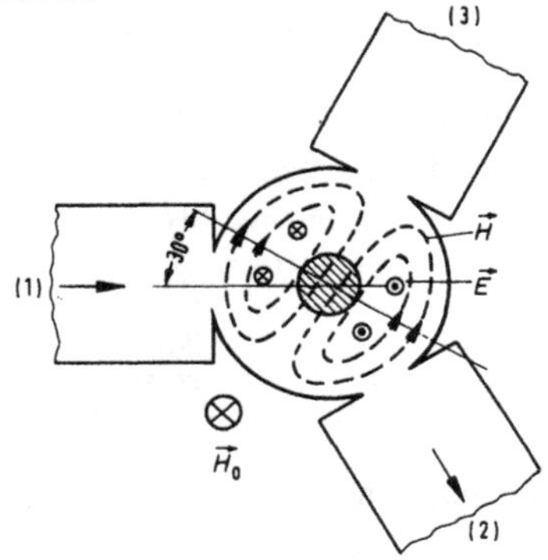


**Fig. 5.17** Y-circulator in waveguide execution

By connecting two triple-gate circulators together, a four-gate circulator can be easily constructed. If one gate of the triple-gate circulator is closed reflection-free, then the arrangement functions as a one-way line.

In Fig. 5.18 some typical application examples of one-way lines and circulators are assembled. In Fig. 5.18a, a generator is decoupled from a receiver with the help of a one-way line and in Fig. 5.18b with a circulator connected as a one-way line; circuit b is attractive for the transmission of high charges because charges reflected from the receiver are absorbed by an adapted terminal resistance. The principle of direction-controlled signal separation with a circulator is represented in Fig. 5.18c by the example of a single-gate amplifier (reflection amplifier); e.g., amplifier with tunnel diode, Impatt or Gunn diode, parametrical amplifier) in the separation of the input and output. In Fig. 5.18d separation is represented for the case of a transmitter and receiver operating at a common antenna. In order to increase the decoupling of transmitted and received parts in directional radio and radar devices, we use additional directional lines (and filters). Figure 5.18e illustrates the construction of a variable phase shifter with the help of a circulator and a removable line short circuit, and Fig. 5.18f shows two circulator circuits operating as variable attenuator pads with adjustable terminal resistance or with fixed terminal resistance and removable line short circuit.

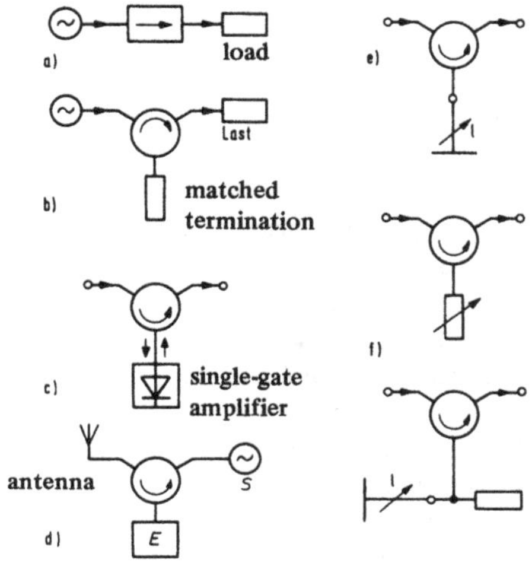


**Fig. 5.18** Application examples of one-way lines and circulators: (a) decoupling with isolator; (b) decoupling with circulator connected as isolator; (c) direction-controlled signal separation with the single-gate amplifier; and (d) with the transceiver; (e) variable phase shifter; (f) variable attenuator pad

## 5.5 YIG-COMPONENTS

Yttrium Iron Garnet ($Y_3Fe_6O_{12}$), abbreviated as YIG, is a ferrite produced by an admixture of yttrium oxide and iron at temperatures over 1200°C. The melting is allowed to solidify into monocrystals which are usually fashioned into small pellets with a very finely polished surface. These YIG pellets are used at their ferrimagnetic resonance as resonators to tune oscillators and filters. YIG monocrystals display a very small spot diameter (approximately 1 MHz) of ferrimagnetic resonance, and therefore a very high unloaded resonator $Q$-factor $Q_0$ up to approximately 10,000, which is determined by the $Q$-factor of the pellet's surface.

For the resonant frequency of the primary mode (110-magnetostatic mode, dipole mode) of the YIG-pellet resonators with a small diameter as compared to the wavelength, such that the propagation effect within the pellet can be disregarded, the following is valid:

$$f_{res} = \frac{1}{2\pi} \cdot \Gamma \cdot (H_0 \overset{(+)}{-} H_a) \,, \quad \Gamma = 0{,}221 \, \frac{\text{MHz}}{\text{A/m}} \tag{5.39}$$

with $H_0$ = outer magnetization constant field

$H_a$ = inner anisotropic field of the crystal (magnitude and sign dependent on the crystallographic orientation of the YIG resonator in the premagnetization field).

The resonant frequency of the YIG element can be linearly changed over a wide range, corresponding to Eq. (5.39), by the intensity of the installed magnetic constant field $H_0$, and therefore the element can be used to electronically tune oscillators and filters.

YIG resonators display several resonances. In addition to the primary (110-) mode, in the case of a non-homogeneous magnetic *HF*-field higher magnetostatic modes occur which can disturb the primary resonance. If we wish to produce a larger frequency-tuning range for the YIG resonator (more octaves possible), then the coupling to undesired higher modes must be kept small. This can be attained with a magnetic constant field $H_0$ in the area of the YIG pellet that is as homogeneous as possible, small dimensions of the YIG element as compared to wavelength, and a large distance from the metal walls (at a small distance there is a decrease of the *Q*-factor).

The temperature dependence of the resonant frequency of the YIG pellet, as a result of the temperature dependent changes of the inner anisotropic field of the crystal, can be reduced by a suitable orientation of the crystal axis in the premagnetization field, and by a thermal compensation of the electromagnet in order to premagnetize. A further decrease of the frequency drift with temperature can be obtained by stabilizing the YIG pellet's temperature through heating.

The lower frequency limit for the use of YIG elements is specified by the saturation magnetization (this means the smallest necessary magnetic field intensity in order to align all the magnetic dipoles), beneath which high losses occur. The upper frequency of the range of application is determined by the necessary strong premagnetization field [Eq. (5.39)] and the diameter of the YIG pellet (should be less than $\lambda/20$ in order to avoid disturbances of the resonance through propagation effects in the pellet). The upper frequency limit lies at approximately 60 GHz. Pure YIG exhibits the smallest losses, although the lower frequency is relatively high with approximately 3.5 GHz. By doping the crystal, usually with gallium, the saturation magnetization can be reduced and thereby the lower frequency can be pushed to approximately 100 MHz. (Reduction of the lower frequency limit is also possible by heating the YIG element.)

The advantages of tuning with YIG resonators are the large tuning range, the good linearity of the tuning ($\omega_{res} = \Gamma \cdot H_0$ proportional to $I$ if no magnetic saturation; linearity is worsened by the reactance of the wiring) and with YIG-tuned oscillators there is a clear spectrum because of the high *Q*-factor.

The coupling of high-frequency signals to the YIG resonator takes place over the magnetic HF-field which should be aligned perpendicular to the premagnetization field so that electron spin precession can be effectively stimulated. YIG resonators permit only a relatively weak HF coupling which is proportional to the saturation magnetization and to the weight of the element.

Figure 5.19a shows the principle of a single-stage YIG band-pass filter. The input signal is coupled to the YIG resonator over a wire loop which concentrates

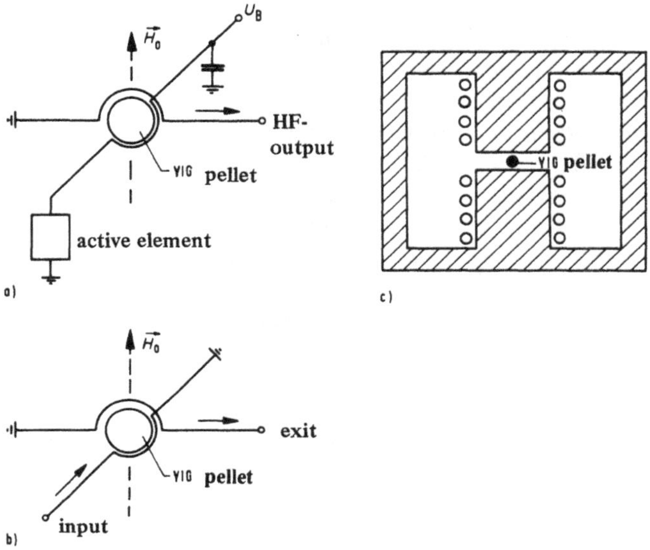


**Fig. 5.19** (a) Diagram of a YIG band-pass filter; (b) YIG-tuned oscillator; (c) magnetic circuit for premagnetization

the magnetic HF-field on the pellet; the decoupling takes place over a second loop. The magnetic HF-field should be adjusted perpendicular to the premagnetization field. The coupling-in and coupling-out loops are arranged perpendicular to one another so that coupling only takes place through the YIG pellet. At the resonant frequency $f_{res} \approx 1/2\pi) \cdot \Gamma \cdot H_0$ of the pellet, premagnetized perpendicular to the HF-field with the static magnetic field intensity $H_0$, the coupling of the signal fed from the input line to the output line takes place over the pellet.

In Fig. 5.19b, the diagram of a YIG-tuned oscillator is represented. YIG-tuned oscillators yield an output signal of high spectral clarity (good stability and low noise) because of the high $Q$-factor of YIG resonators. In comparison with variable reactor-tuned oscillators, with the YIG-tuning we obtain the advantages of good linearity, a very wide tuning range (up to many octaves), and the disadvantages of a low tuning velocity as a result of the magnetic circuit inertia, a higher power dissipation, and a heavier weight of the oscillator (due to the electromagnet). Additionally, the YIG tuning is only applicable up to approximately 60 GHz because of the necessarily strong magnetic field.

By coupling the YIG resonator to the circularly polarized magnetic HF-fields, we can attain a non-reciprocal attenuation behavior which is used to produce isolators and circulators.

An additional group of YIG components are the MSW elements (MSW, or magnetostatic surface wave) which are used for tasks similar to those of SAW elements (SAW, or surface acoustic wave).

*SAW elements* are not ferrite components; they should, however, first be illustrated in this connection. With them *acoustic surface waves* propagate on the surface of a suitable piezoelectric substrate (quartz, lithium-niobate, $LiNbO_3$) with acoustic velocities of approximately $3.1 \cdot 10^3$–$3.4 \cdot 10^3$ m/s. The waves are excited by comb-shaped electrodes stacked on top of each other, so-called interdigital transformers. This transforms the input HF energy into acoustic energy of the lattice vibrations of the substrate, and back into HF energy at the output of the element. The SAW element functions, thus, as a delay line and can be used for signal processing tasks (e.g., formation of the Fourier transform), as band-pass filters, or as pulse-compression or optimal filters (matched filters) in radar engineering. It is often installed as a frequency-selective element (operating as a delay element or resonator) in the backward coupling path of an amplifier, whereby the circuit then functions as an oscillator. Such SAW oscillations exhibit high frequency stability (suitable as a reference oscillator) and a good spectral clarity. They have small size and light weight. These oscillators can be constructed in hybrid technology (see Ch. 4); the SAW chip is quite suitable for hybrid circuit integration as a planar element. They are also constructed as monolithic circuits, in which the SAW element and the semiconductor amplifier are produced on the same GaAs substrate. Today, SAW oscillators are being produced for frequencies within the range of approximately 20 MHz to 3 GHz.

In order to complete the tasks mentioned, *MSW elements* are also used operating at frequencies of approximately 1 GHz to above 10 GHz. With these, we use *magnetostatic volume-* or *surface waves* in ferrite crystals. The waves are excited by metal strips or irregular-shaped transformers. A thin epitaxial YIG film will serve as a propagation medium, which is mounted in planar geometry on the gadolinium-gallium-garnet (abbreviated as GGG) and is premagnetized with an outer magnetic field ($H_0$). The mechanism of magnetostatic wave propagation is not based on the excitation of lattice vibrations, as was the case with the SAW elements, but rather upon the ferrimagnetic coupling. In this connection, the magnetization vector precesses around the $H_0$ direction, as discussed in Ch. 5. The propagation velocity is greater than in the case of acoustic surface waves and is approximately $5 \cdot 10^5$ m/s.

Depending on the direction of the premagnetization field with reference to the wave propagation direction, the following waves are excited: surface waves at $H_0$, perpendicular to the direction of propagation and in the direction of the YIG layer; forward-volume waves at $H_0$, perpendicular to the direction of propagation and the YIG layer; or backward-volume waves at $H_0$, opposite the direction of propagation and in the direction of the YIG layer.

## SUMMARY

To summarize, in this book we first illustrated the propagation of electromagnetic fields in free space in the form of plane waves. Then waveguides for conducting electromagnetic waves were examined and then we discussed the wave modes which are obtained within the boundaries of the limit conditions to be satisfied from the fields to the conductor boundaries. Double-conductor arrangements, such as, for example, the Lecher- and the coaxial transmission line, conduct *H*-waves, for which (for $\kappa_{\text{conductor}} \rightarrow \infty$) the electrical and magnetic fields are directed perpendicular to the wave propagation direction as in the case of plane waves. In the case of waveguide waves, either the magnetic field possesses a component in the direction of propagation (*H*-waves), or the electrical field does so (*E*-waves). Then, various stripline type models were considered, in which the conductor paths are mounted on a dielectric carrier plate. Dielectric circuits are also described, which transport the field energy in the form of surface waves.

In Chapter 2 the straggling parameters (S-parameters) were introduced, which help to describe microwave components and circuits (*N*-gates). They are defined with the help of incident and reflected (power) waves at the gates of the circuit element. In using S-parameters, the resulting pole frequency response location representations of the amplification and the stability criteria (e.g., for solid-state amplifiers) was also discussed.

A great number of microwave components can be realized with the help of line sections. Thus, we use open or short circuited line sections of a determined length at the terminations as reactances, or we use, for example, sudden changes in the line oscillation resistance by variation of the strip lines conductor width in the construction of accommodating connections and filter circuits. In waveguide technology, metal diaphragms, pins, or dielectric inserted in the hollow waveguide often serve for such circuits. Active resistances can be realized by the installation of absorbing material in the line or as a circuit termination. With coupled lines, directional couplers and filters can be constructed. Still to be mentioned is the treatment of line bypasses (hybrids), transitions, waveguide rotating joints, and interrupters. Circuit components in waveguide technology are described in Chapter 3 and those in stripline technology, in Chapter 4. Stripline technology is well suited to the construction of microwave integrated circuits (MIC). Such circuits are constructed as either hybrid or monolithic circuits in thin-film or thick-film technology. In Chapter 4 distributed and concentrated stripline operating components are examined.

Circuit components which work with the insertion of ferrite material into the line are dealt with in Chapter 5. With these components, we often exploit the fact that the electron spin precession is only excited by a clockwise circularly polarized wave in order to realize a non-reciprocal transmission behavior with one-way lines and circulators. Attenuator pads and phase shifters are also constructed with ferrites. Also illustrated are YIG resonators, which are used for the tuning of filters and oscillators, and MSW elements that work with magnetostatic surface waves. These are used as delay lines, in the treatment of signals, and as filters and backward-coupling networks in oscillators.

## Bibliography


1.Collin, R.E., *Grundlagen der Mikrowellentechnik,* 1973, VEB Verlag Technik, Berlin.

2.Grivet, P., *Microwave Circuits and Amplifiers,* Vol. 2, 1976, Academic Press, London.

3.Harvey, A.F., *Microwave Engineering,* 1963, Academic Press, London.

4.Helszajn, J., *Passive and Active Microwave Components,* 1978, John Wiley & Sons, New York.

5.Howe, H., *Stripline Circuit Design,* 1974, Artech House, Dedham.

6.Janssen, A., *Hohleiter und Streifenleiter,* 1977, Dr. A. Huthig Verlag, Heidelberg.

7.Jansen, R.H, *Spezialproblme der Mikrowellentechnik,* Duisburg, 1979.

8.Megla, G., *Dezimeterwellentechnik,* 1962, Berliner Union, Stuttgart.

9.Marcuvitz, N., *Waveguide Handbook,* 1951, McGraw-Hill, New York.

10.Meinke, H., Gundlach, F.W., *Taschenbuch der Hochfrequenztechnik,* 1968, Springer, Berlin.

11.Meyer, H. Gundlach, F.W., *Physikalische Grundlagen der Hochfrequenztechnik,* 1969, Fr. Vieweg & Sohn, Braunschweg.

12.Michel, H.J., *Zweitor-Analyze mit Leistungswellen,* 1981, B.G. Teubner, Stuttgart.

13.Rint, C., *Handbuch fur Hochfrequenz- und Elektro-Techniker,* Vol. 5, 1981, Dr. A. Huthig Verlag Heidelberg.

14.Saad, T.S., *Microwave Engineers Handbook,* 2 vol. 1971, Artech House, Dedham.

15.Schelkunoff, S.A., *Electromagnetic Waves,* 1943, Van Nostrand, New York.

16.Severin, H, *Leitungen fur Oberflachenwellen im Hochstfrequenzgebeit,* Philips Technical Review, No. p. 26 (1965), h. 1/2, s. 24-40.

17.Wolf, I, *Einfuhrung in die Mikrostrip-Leitungstechnik,* Verlag H. Wolf, Aachen.

18.Young, L., *Advances in Microwaves,* Volumes 1-8, 1966-1974, Academic Press, London.

19.Zinke, O., Brunswig, H., *Lehrbuch der Hochfrequenztechnik, Vol. I, 1973, Vol. III, 1974, Springer, Berlin.*

20.*Fradin, A.S., Microwave Antennas,* 1961, Pergamon Press, Oxford.
21.Hansen, R.C., *Microwave Scanning Antennas,* Vol. 1, 1964, Vol. 2, 3., 1966, Academic Press, New York.
22.Heilmann, A., *Antennen III,* 1970, BI Hochschultaschenbucher, Mannheim.
23.Kuhn, R., *Mikrowellenantennen,* 1964, VEB Verlag Technik, Berlin.
24.NTG-Entwurf 1301, *Begriffe aus dem Geibiet der Antennen*, NTZ (1969), No. 6, p. 325-330.
25.Silver, S., *Microwave Antenna Theory and Design,* 1949, McGraw-Hill, New York.
26.Trentini, G.V., *Ubersicht der heute in der Technik verwendeten stark bundelnden Mikrowellenantennen, Frequenz* 29 (1975) Vol. 1, No. 6, p. 158-164, Vol. 2, No. 7, p. 192-199.
27.Bosch, B.G., Engelmann, R.W., *Gunn-effect Electronics,* 1975, Pitman Publishing.
28.Collins, G.B., *Microwave Magnetrons,* 1948, McGraw-Hill, New York.
29.Flyagin, V.A., *et. al.,* The Gyrotron, *"IEEE Transactions on Microwave Theory and Techniques,"* Vol. MTT-25, (1977), p. 6.
30.Frey, J., *Microwave Integrated Circuits,* 1977, Artech House, Dedham.
31.Graham, E.D., Gwyn, C.W., *Microwave Transistors,* 1975, Artech House, Dedham.
32.Kowalenko, W.F., *Mikrowellenrohren,* 1957, Porta Verlag, Munchen.
33.Liechti, C.A., Microwave field-effect transistors, *"IEEE Transactions on Microwave Theory and Techniques,"* Vol. MTT-24 (1976), No. 6, p. 279-300.
34.Renz, E., *PIN- und Schottkydioden,* 1975, Dr. A. Huthig Verlag, Heidelberg.
35.Rint, D., *Handbuch fur Hockfrequenz- und Elektro-Techniker,* Vol. 3, 1979, Dr. A. Huthig Verlag, Heidelberg.
36.Tholl, H., *Bauelemente der Halbleiterelektronik,* Vol. 1, 1976, B.G. Teubner Verlag, Stuttgart.
37.Unger, H.G., Harth, W., *Hochfrequenz-Halbleiterelektronik,* 1972, Hirzel Verlag, Stuttgart.

***Additional References***

1.Frey, J., Bhasin, K., *Microwave Integrated Circuits,* 2nd ed., 1985, Artech House, Dedham.
2.Gardiol, F., *Introduction to Microwaves,* 1984, Artech House, Dedham.
3.Hoffmann, R.K., *Handbook of Microwave Integrated Circuits,* 1985, Artech House, Dedham.
4.Pehl, E., *Mikrowellentechnik,* 2 vol., 1984, Dr. Alfred Huthig Verlag, Heidelberg.

www.ingramcontent.com/pod-product-compliance
Lightning Source LLC
Chambersburg PA
CBHW021430180326
41458CB00001B/207
* 9 7 8 0 8 9 0 0 6 1 6 4 0 *